Fundamentals of Soil Chemistry

NIPA® GENX ELECTRONIC RESOURCES & SOLUTIONS P. LTD.
New Delhi-110 034

About the Author

Dr. Dibyendu Mukhopadhyay is a Professor of Soil Science and Agricultural Chemistry in Uttar Banga Krishi Viswavidyalaya. Dr. Mukhopadhyay is involved in teaching Soil Science and on research activities over 22 years. He received the BSc degree (1994), MSc Degree (1996) and Ph.D. degree (2002) in Agricultural Chemistry and Soil Science from Bidhan Chandra Krishi Viswavidyalaya. Dr. Mukhopadhyay started his career as Lecturer in the discipline of Soil Science and Agricultural Chemistry at Uttar Banga Krishi Viswavidyalaya in 2001. So far, fifteen students in MSc. and four in Ph.D. have been conferred their degree of Uttar Banga Krishi Viswavidyalaya under his direct supervision and guidance. He has published fifty research papers in National and International journal of repute till date. Dr. Mukhopadhyay is a fellow of the Society of Tropical Agriculture (New Delhi) and also has discharged the Editorial and as Reviewer for a number of journals. He was the Head, Department of Soil Science and Agricultural Chemistry and has been entrusted with higher responsibilities for managing the Agricultural Education as Dean, Faculty of Agriculture, Uttar Banga Krishi Viswavidyalaya, Pundibari, West Bengal, India.

Fundamentals of Soil Chemistry

Dibyendu Mukhopadhyay M.Sc. (Ag.), Ph.D.
Professor
Soil Science and Agricultural Chemistry
Uttar Banga Krishi Viswavidyalaya
Pundibari, CoochBehar, West Bengal, India

NIPA® GENX ELECTRONIC RESOURCES & SOLUTIONS P. LTD.
New Delhi-110 034

NIPA® GENX ELECTRONIC RESOURCES & SOLUTIONS P. LTD.

101,103, Vikas Surya Plaza, CU Block
L.S.C. Market, Pitam Pura, New Delhi-110 034
Ph : +91-11-43860225, Mob.: +91 9717133558, 9540816132
E-mail: newindiapublishingagency@gmail.com
Website: www.nipaersources.com

Print ISBN: 978-93-58872-32-3
ebook ISBN: 978-93-58877-95-3

Composed and Designed by NIPA®

Dedicated to
My Parents

PROF. SAROJ KUMAR SANYAL, M.Sc. (IARI)
Ph.D. (Cambridge) FISSS, FNAAS, FAScT
Former Vice-Chancellor
Bidhan Chandra Krishi Viswavidyalaya
Mohanpur-741 252, District-Nadia. West Bengal
Former Adjunct Professor
Indian Agricultural Research Institute, New Delhi
Former Chairman, Research Advisory Committee
NBSS & LUP (ICAR), Nagpur & CRIJAF (ICAR)
Barrackpore, West Bengal
Former Visiting Scientist
International Rice Research Institute
The Philippines President, Section of Agriculture
& Forestry Sciences, 92nd Session of the Indian Science
Congree Association (2004-2005)
Member, QRT, ICAR-NRRI, Cuttack
Ref. No. P/784/ 2024-2025

Residence
Flat A1 (Ground Floor)
Saptaparni
58/3, Ballygunge Circular Road
Kolkata-700 019
Telephone: 033-24544694
Mobile: 098302 47072
sarojsanyal@yahoo.co.in
sksanyalnaip@gmail.com
Dated, May 14, 2024

Foreword

I take immense pleasure in writing the Foreword for the book entitled, 'Fundamentals of soil chemistry', written by Dr. Dibyendu Mukhopadhyay, Professor of Soil Science and Agricultural Chemistry, Uttar Banga Krishi Viswavidyalaya, Pundibari, Coochbehar. Soil, the precious natural gift of nature and the habitat of so many biotic and abiotic substances undergoes complex processes pertaining to its formation. The concept of basic principles of soil chemistry is essential to understand the nutrient dynamics in soil-plant systems affecting the sustainable production in the ecosystem. Soil chemistry is considered to be the stepping stone for illustrating the environmental chemistry and green chemistry and is equally important for other branches of soil science. The book, 'Fundamentals of Soil Chemistry', is a good attempt taken by Dr. Mukhopadhyay. The author has given his fullest endeavour in this book consisting of ten chapters covering various areas on soil chemistry with updated information. The content of each chapter has been arranged systematically with discussions and diagrams considering the need of the students, researchers and other cognate disciplines. The basic principles of soil chemistry,the soil colloidal properties, the concept of ion exchange phenomenon and its significance in plant nutrition,the function of acid and alkaline soils and their reclamation processes,the adsorption and desorption of nutrients in soil and its implications towards fertilizer application have been discussed in different chapters of the book. The concept on chemistry and implication of submerged soils and stability of organo-metal complexation in soil are some of the lines of interests in the field of soil science that have been included in the book. The book is written in a very simple and lucid language. The author has also figured out relevant questions at the end of each chapter which will be helpful for the students. I believe, this book will be a source of information for those who are engaged in research and teaching with its novelty to keep pace with the rapid advancement in the science of chemistry.

Saroj Kumar Sanyal

Preface

The book, **'Fundamentals of Soil Chemistry'**, is meant for the students of Soil Science in particular and for those who are studying agriculture in general. The book is written in a simple language to understand the basics of Soil Chemistry and its applications. This book will be useful both for the undergraduate and post graduate students. The book comprises of ten chapters on fundamental aspects of soil chemistry. The chapter 1 deals with the basic principles of chemistry while chapter 2 is intendent to know the composition of soil. The elements in plant nutrition and soil organic matter are discussed in chapter 3 and 4 respectively. The soil colloidal properties, the structure of silicates and non-silicate minerals, identification of clay minerals as well as charge characteristics of clay are presented in chapter 5. The concept of ion exchange phenomena and its significance in plant nutrition are addressed in chapter 6. The function of acid and alkaline soils and their reclamation processes are discussed in chapter 7. The effect of soil reaction on plant growth has also been discussed in this chapter. The adsorption and desorption of nutrients in soil and its implication towards fertilizer application are dealt in chapter 8. The chemistry and implication of submerged soils are discussed in chapter 9. The organo-metal complexation in soil and the nature of stability of the complexes are discussed in chapter 10.

I deeply acknowledge the constructive suggestions and advise received from Dr. S.K. Sanyal, Professor of Agricultural Chemistry and Soil Science and Former Vice-Chancellor of Bidhan Chandra Krishi ViswaVidyalaya (BCKV), Mohanpur,West Bengal during the preparation of the manuscript. Sincere thanks are also extended to Dr. D.K.Das, Dr. S.K.Pal and Dr. P.K.Mani of Bidhan Chandra Krishi ViswaVidyalaya, Mohanpur and also to my colleagues Dr. A.K.Sinha, Dr. G.C.Banik, Dr. A.Tamang, Dr. S. Deb, Dr. K. Karmakar, Dr. A.Choudhury and Dr. A.K. Saha, students and staffs of Uttar Banga Krishi Viswavidyalaya, Pundibari for their continuous support and encouragement. The sources of figures and tables which have been reproduced from different journals and books have been duly acknowledged.

I wish to express my heartful appreciation to my wife Sampa, my daughter Aishi and son Ayaan for their constant encouragement and support and also from my sisters Soma and Mampi without which it would have been impossible to complete the books. Last but not the least, the blessings of my parents in disguise gave me strength to complete the book.

Hope, this book will help the students to understand the fundamentals of soil chemistry and will guide them to undertake research work in this subject. I would appreciate inviting constructive criticism and suggestion of the book, which would provide directions in improving the future editions.

Dibyendu Mukhopadhyay

Date: May 14, 2024
Pundibari

(Dibyendu Mukhopadhyay)

Contents

1

Principles of Chemistry

The knowledge on fundamental principle of chemistry is important to understand and examine the mechanisms of reactions to form products. The chemistry of soil processes will also help understanding the correlation between the basic and applied sciences for achieving the sustainable ecosystem in the environment. The formation and stability of bonds, thermodynamic principles, radioactivity, coordination number etc. are essential to solve the on-going interactions in nature. The structure of atomic models, valence and equivalent weight, creating elementary particles, quarks, neutrinos, bosons are some of the important arena where fundamental study and research can be explored for the basic and strategic ideas. Hence, it is very much essential to apprehend the concept of physical chemistry, soil chemistry and geochemistry to correlate with the definitions of chemistry for understanding the source/sink properties of soils for nutrient adsorption and release characteristics.

1.1 Chemical Bonds

It is established from the Periodic table that, the arrangement of electrons in atoms of the inert gases is more stable than in other elements. Again by the exchange or sharing of electrons with the other atoms, the more reactive or less stable elements can be achieved as is found in the inert gases. It is by this exchange or sharing of electrons which can form the chemical bonds to bind the atoms together. An atom can achieve the inert gas electron arrangement by i) loosing one or more electrons to another atom; ii) gaining one or more electrons from another atoms; iii) sharing one or more electrons with another atom.

Hydrogen

Carbon

Water

Ethylene

Acetylene

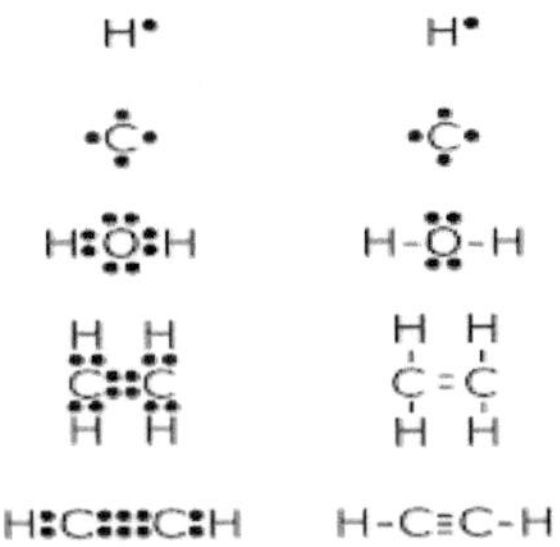

1.2 Ionic or Electrovalent Bonds

Ionic bonds are formed by the exchange of electrons between atoms. The transfer of electrons take place in such a way that the ions formed have the stable electronic arrangement of an inert gas. Thus, when a metal reacts with a non-metal , transfer of electrons take place from metal atoms to the non-metal atoms and an ionic bond is formed. A strong force of attraction is developed between the oppositely charged ions in an ionic bond. When sodium reacts with chlorine, it transfer its one outermost electron to the chlorine atom. In potassium chloride , the electronic configuration of both potassium ion as well as chloride ion (2,8,8), results that of inert gas argon (Figure 1.1). Example: Formation of NaCl, KCl.

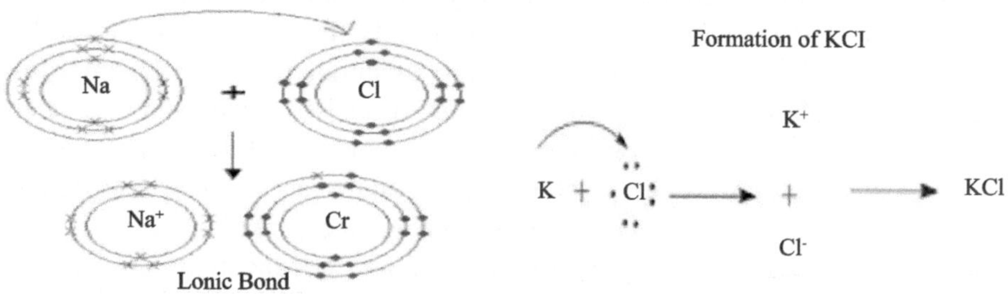

Figure 1.1. Electronic configuration of NaCl and KCl

Potassium (K) has one electron in its outer shell or orbit. Each 'K' atom requires to exclude one electron to form a charged atom or ion K^+ to achieve the inert gas's structure of Argon (Ar). On the otherhand, the Cl contains seven electrons in its outer shell which requires one additional electron to gain an inert gas structure. This will result in a charged atom Cl^- resembling the structure of Ar. As a result of this electron exchange, both atoms become charged and the positive K ion (Cation) will attract the negative Cl^- (anion). Thus, attraction between the oppositely charged ions (i.e., K^+, Cl^-) form the ionic bond (Figure 1.1).

1.3 Covalent Bonds

This type of bonds are formed by sharing of outermost electrons between atoms. The sharing of electrons take place in such a way that the resulting molecule gets the stable electronic arrangement of an inert gas. When a non-metal combines with another non-metal, sharing of electron take place between these atoms and a covalent bond is formed, e.g. HCl. Covalent bonds are of three types;

i) Single covalent bond, ii) Double covalent bond, iii) Triple covalent bond. **Example:** Formation of gaseous element of Cl.

It is known that, Cl atom needs one electron to achieve the stable electronic configuration of Argon (Ar). However, in Cl, instead of gaining an electron outright as in the ionic bond of KCl, each atom shares one of its seven electrons with another Cl atoms (Figure 1.2).

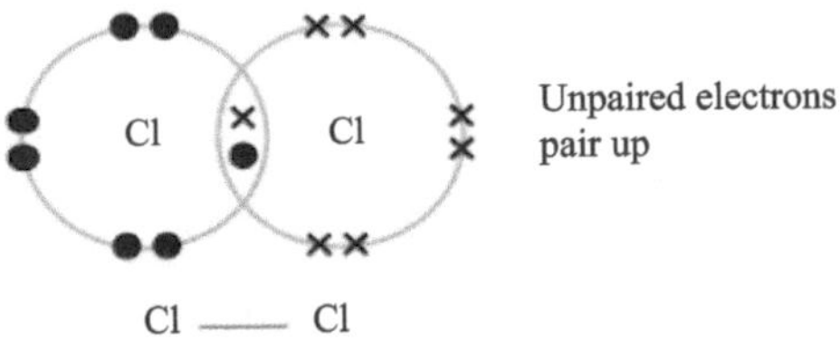

Single bond formation

Figure 1.2. Formation of single bond in chlorine

Thus, each atom has a share in eight outer electrons (an octet). Shared electrons hold the two atoms together in a single covalent bond to form a diatomic molecule (i.e.,Cl_2). A double covalent bond is formed by sharing of two pairs of electrons between the atoms. i.e., four electrons are involved between the two atoms (Figure 1.3). A double bond is a combination of two single bonds. Examples : O_2,N_2,CO_2,H_2O,CH_4,C_4H_{10},$CHCl_3$.

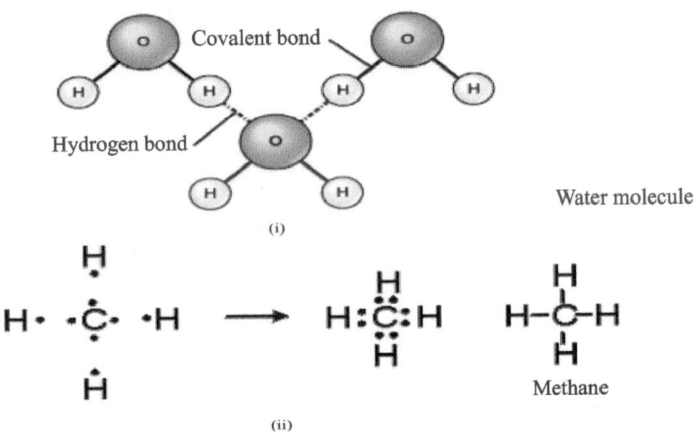

Figure 1.3. Formation of Hydrogen bond (i) and Single bond (ii)

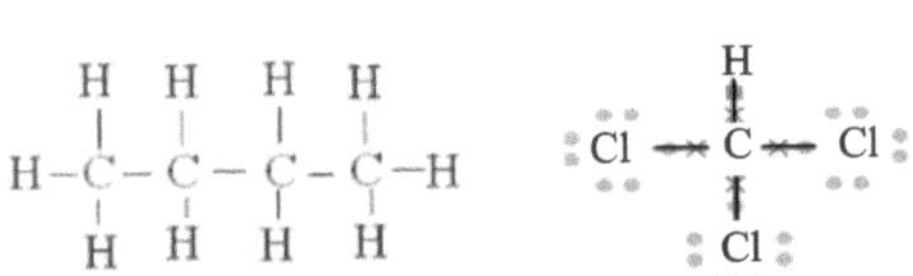

Butane Chloroform

Carbon dioxide contains two double bonds. Each double bond is comprised of one sigma bond and one π bond.

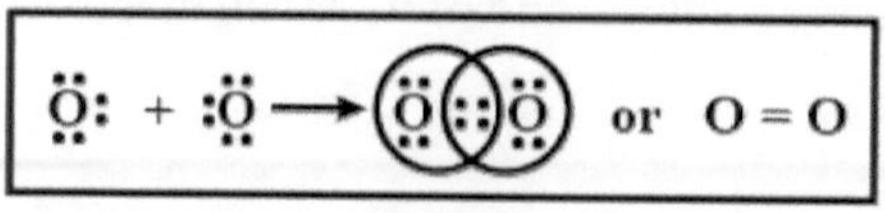

Double bond, sharing of four electrons

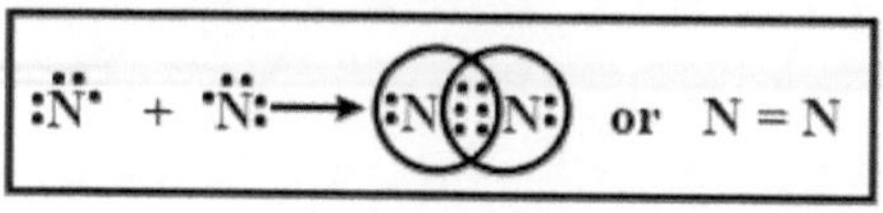

Triple bond, sharing of six electrons

A triple bond is formed by the sharing of three pairs of electrons between two atoms. Hence, there is involvement of six electrons between two atoms, each atom contributing three electrons for sharing.

H:C:::C:H or H:C≡C:H

Acetylene

1.4 Ionic and Covalent Compound

The compounds of metals and halogens (the metal halides)are predominantly ionic, e.g. NaCl,KCl,NaF,KBr,$MgCl_2$ etc. but, there are some atoms consisting of a number of atoms (poly atomic) covalently bonded together.

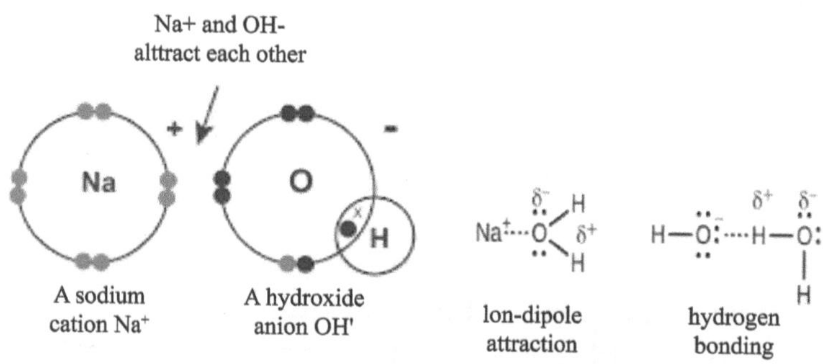

Figure 1.4. Formation of Ionic and Co-valent compounds

Here, NaOH is made up of sodium (Na^+) and hydroxyl ions (OH^-), held together by ionic bonds. Since the oxygen and hydroxyl atoms are firmly bonded together, the hydrogen ion does not normally breakdown in chemical reactions (Figure 1.4).

Examples of Polyatomic ions: NO_3^-, NH_4^+,SO_4^{--},CO_3^{--},$H_2PO_4^-$,HPO_4^{--},PO_4^{---}

1.5 Polarised Covalent Bonds and Hydrogen Bonds

In a covalent bonds between atoms of different electronegativity, the shared electrons tend to be displaced towards the more electronegative atoms giving a polarised covalent bonds. A polar bond is a covalent bond in which there is a separation of charge between one end and the other.

Example : The OH^- bond in a water molecules having the electronegativities of H and O are 2.1 and 3.5 respectively, where, the bonding electrons are slightly displaced towards the O-atom resulting a partial separation of charge.

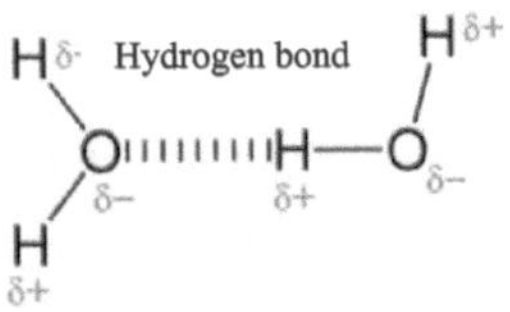

In hydrogen bond, a weak bond exists between two molecules resulting from an electrostatic attraction between a proton in one molecule and an electronegative atom in the other. Such a bond is although weaker than an ionic bond or covalent bond but relatively stronger than the van der Waals force of attraction. Because of the polarization of the O-H bond in water molecules, the partial negative charge on the O - atom of each molecule will be attracted to the partial positive charge on the nearest neighbouring water molecule giving a network of hydrogen bonds between the molecules of the liquid water. A hydrogen bond is formed between a hydrogen atom (carrying a fractional positive charge) or one molecule and a more electronegative atom (not necessarily oxygen carrying fractional negative charge) or another molecule.

Other Bonds

1.6 Dative Bond

Here, one electron is contributed by each partner but situations are there where both electron of a single covalent bonds are contributed by one of the participating atoms. This type of bond is called a dative bond or dative covalent bond.

Example :

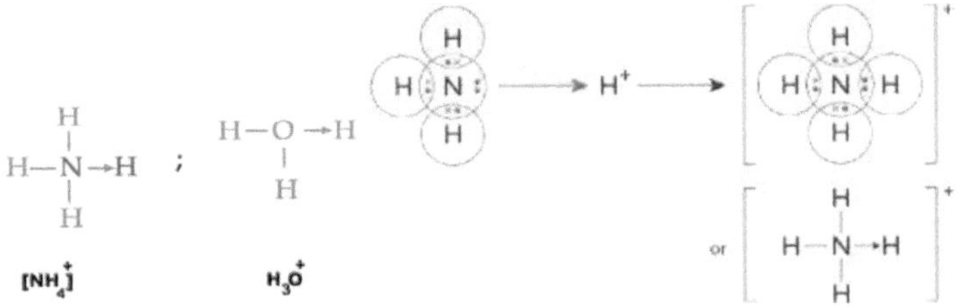

In aqueous solutions, an ammonia molecule will form a datine covalent bond with a portion by contributing a pair of electrons to give an ammonium ion.

Thus, the N can maintain its octate while the hydrogen can attain the electronic structure of Helium. The formation of dative bonds are most frequent in the formation of complex ions, which is formed by the simple ions through the covalent bonding with the other ions, molecules or atoms. These electron-dominated ions or groups or molecules contributing pair of electrons to the central metallic ions are called ligands.

1.7 Metallic Bond

In general, the metals have one or two electrons in the outermost shell of their atoms. These metals are highly electropositive where the electrons are not tightly bound to the atoms. In the solid metals, atoms are closely packed , while their loss of outer electrons (one or two per atom) are contributed to the shared pull of electrons. The high electrical conductivity in the metals is possible due to the looseness and mobility of the cloud of electrons. The atoms that the electrons leave behind become positive ions and on interaction between these ions and valence electrons gives rise to the cohesive or binding force holding the metallic crystal together. The metallic bonds are strong and require lot of energy to break, causing high melting and boiling points.

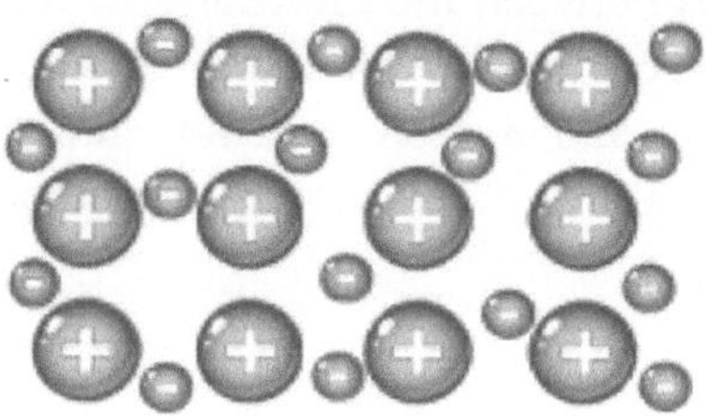

Metallic Bond

1.8 Shape of the Molecules

The shape of a molecule is associated with the molecular geometry of the atoms in a three dimensional arrangement including the shape of the molecule,bond length, bond angles, torsional angles and geometric parameters governing the position of each atom (Figure 1.5). This shape also depends upon the preferred spatial orientation of covalent bond to atoms having two or more bonding partners.

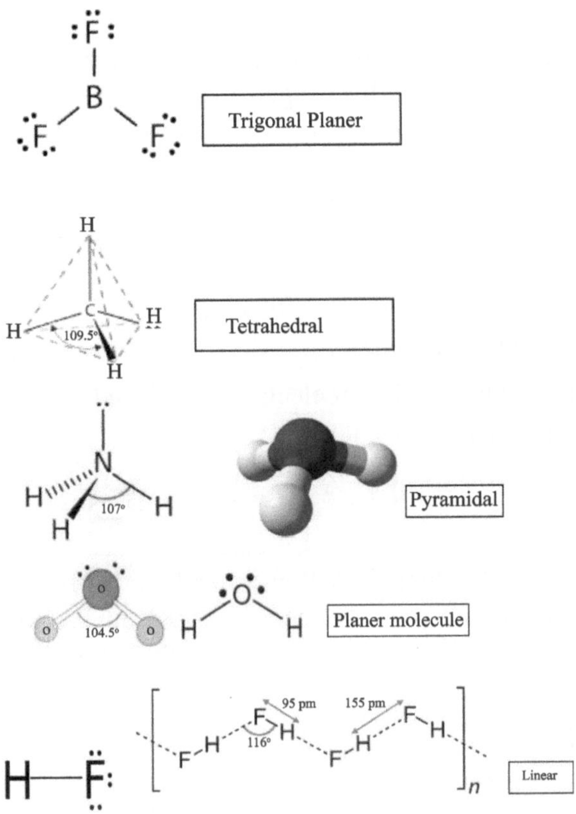

Figure 1.5. Shape of molecules with different bonds

Hydrocarbon

A compound made up of hydrogen and carbon is called hydrocarbon. Methane, ethane, ethene and ethyne are hydrocarbons made up of hydrogen and carbon. The petroleum/crude oil or natural gas also contain hydrocarbon. Among the two types of hydrocarbon (saturated and unsaturated), Methane, ethane are saturated and ethene, ethyne are unsaturated hydrocarbons. The saturated hydrocarbons (alkanes) are chemically not very reactive.

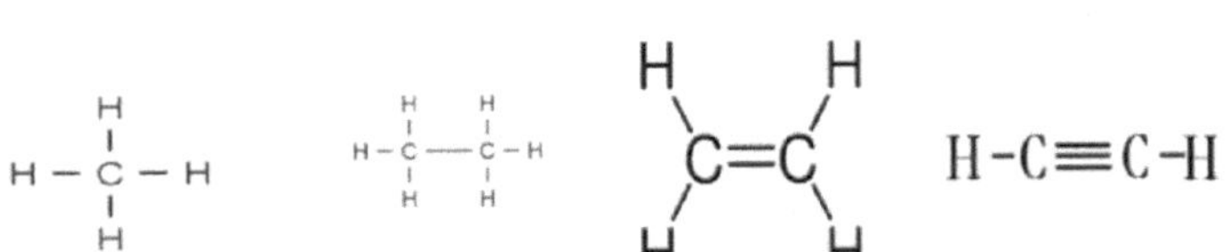

Methane Ethane Ethene Ethyne

1.9 Equivalent Weight

It is the number of parts by weight of an element which will combine with 8 parts by weight of oxygen, or 1.008 parts by weight of hydrogen or will displace these amounts from their compounds. Thus, 14/3 g of nitrogen are known to combine with 1 g of hydrogen to form ammonia; Again, 8 g of oxygen combine with 1 g of hydrogen to yield water. The equivalent weight of an element is the mass which combines with or displaces 1.008g of hydrogen or 8.0 g of oxygen or 35.5 g of chloride. The equivalent weight of an element is its gram atomic weight divided by its valence (combining power).

Equivalent weight = Atomic weight (g)/valency

Equivalent weight = Molecular weight /number of electron transferred

$$Fe^{+2} \rightarrow Fe^{+3} + e^{-} \text{ Eqw} = 1/1 \text{ Fe}$$

$$As^{+3} \rightarrow As^{+5} + 2e^{-} \text{ Eqw} = ½ \text{ As}$$

1.9.1 Avogadro's Law

Avogadro's hypotheses states, 'Equal volumes of all gases and vapour under the same conditions of temperature and pressure, contain equal number of molecules'. The law is approximately valid for real gases at sufficiently low pressures and high temperatures.

1.9.2 Avogadro's Constant(N)

Number of molecules per mole of substance

N = 6.02278 x10^{23} per mole

The Avogadro's constant is rather the proportionality factor relating to the number of constituent particles in a sample with the amount of substance in that sample.

1.9.3 Molecular Weight

The molecular weight is the mass of one mole of a substance. The molecular weight of any gas or vapour is equal to twice its density assuming that the hydrogen molecule is diatomic and its atomic weight is equal to one. Usually, the unit used for this is gram per mole. The sum of the atomic masses of all atoms in a molecule based on a scale in which the atomic masses of hydrogen,carbon,nitrogen and oxygen are 1,12,14 and 16 respectively.

1.9.4 Atomic Weight and Atomic Mass

It is expressed as the ratio of the weight of an atom of the element to the weight of an atom of a reference element like oxygen (O=16).The atomic mass of an element is the average mass of the atom of an element measured in atomic mass unit (amu) and is associated with the number of neutrons and protons that are present in a particular nucleus of an element.

Physical atomic weight scale : Chemical atomic weight scale x 1.00027

The exact atomic weight of an element can be determined as :

Atomic weight = Equivalent weight x Valency

Gram atomic weight = atomic weight in grams. Hence, one gram atom of carbon = one gram atomic weight of carbon = 12 grams. Some of the characteristics of the elementary particles are given in Table 1.1.

Table 1.1. Characteristics of Elementary Particles

Particle	**Symbol**	**Mass (a.m.u.)***	**Charge (e.s.u.x10^{10})**	**Relative charge**
Electron	e^-, β^-	0.0005486	-4.8029	-e
Proton	$^1H_1$, p	1.00757	+4.8029	+e
Neutron	$^1n_0$, n	1.00893	0	0
Positron	e_1^+, β^+	0.0005486	+4.8029	+e
Antiproton	p^-	1.00757	-4.8029	-e
Neutrino	N	0.005	0	0
Anti-neutrino	ν'	0.005	0	0
Meson	$\pi \pm$	0.156	± 4.8029	±e
	$\mu \pm$	0.118	±4.8029	± e

*a.m.u is the atomic mass unit as expressed on physical atomic weight scale;
1 a.m.u. = 1.6603 x 10^{-24} g

Atomic number : The atomic number of an element is equal to the number of positive charges (protons) on and the number of electrons distributed around the nucleus as is roughly equal to half the atomic weight. The atomic number can determine the place of a particular atom in a periodic table. Hence, the number of proton determines the element's atomic number distinguishing one element from the other, while, an element's mass number is determined by the sum of the number of protons and the number of neutrons.

1.9.5 Atomic Structure

It refers to the structure of an atom comprising a neucleus in which the positively charged protons and neutral neutrons are present (Figure 1.6). The negatively charged particles (electrons) move around the center of the nucleus.

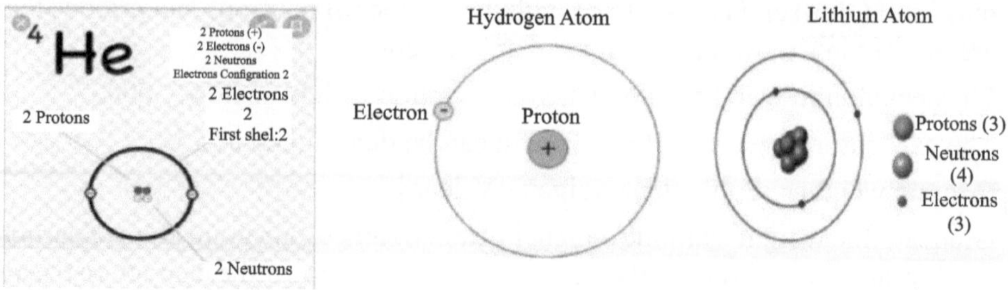

Figure 1.6. Formation of atomic structure

1.9.6 Important Functionalities in Chemical Reactions

Normality (N) : The number of equivalent weights of a substance dissolved in 1 litre (L) of a solution is called normality (N) or strength of the solution. It is a measure of concentration equal to the gram equivalent weight per litre of solution. The solute's role in the reaction determines the solution's normality.

e.g., to make 1 N solution of NaOH , 40 g of NaOH are to be dissolved in 1 L. Similarly, for a 0.1N solution of NaOH, 4 g of NaOH is to be dissolved in one litre of a solution.

Molarity (M) : The number of moles of a substance per litre(L) of solution is called molarity (M).

Molality (m) : The number of moles of a substance dissolved per kilogram of water is called molality (m). e.g. Molecular weight of H_2SO_4 = 2+32+64 =98;1 L weighs 1820g

Hence, molarity of H_2SO_4 = 1820/98=18M

Formality (f) : The number of moles of a substance per kg of solution is called formality(f).

Mole fraction: The relative amount of a component in a mixture is expressed as a fraction of total amounts of all the components , the amount of every component being measured in gm-molecules.

e.g. in a solution, n_1 gm-moles of A and n_2 gm-moles of B will have total amount of (n_1+n_2) gm-moles.

Therefore , the mole fraction of component A = $n_1/(n_1+n_2) = x_1$ (say) and mole fraction of component B = $n_2/(n_1+n_2) = x_2$ (say)

Hence, $x_1+x_2 = 1$

If there are i-components present in the solution, the sum of the mol-fractions of all the components would be unity.

$\sum x_i = x_1 + x_2 + x_3 + \ldots\ldots + x_n = 1$

1.10 Oxidation –Reduction (Redox) Potential

The chemical reaction in which electrons are transferred from a donor to an accepter is called the redox-reaction. Or in other words, the single electrochemical properties that serves to distinguish a submerged soil from a well- drained soil is its redox potential. There is change in the oxidation states of ions (Table 1.2) and also in free energy (ΔG) due to chemical reactions. In a reduction process , it can be read as :

$$Ox + ne \leftrightarrow Red$$

$$\Delta G = \Delta G^0 + RT\ln (Red)/(Ox)$$

where, (Red) and (Ox) are the activities of the reduced and oxidised species and ΔG^0 is the free energy change when the activities are unity.

again, $\Delta G = -nEF$

According to Nernst equation,

$E=E_0 + RT/nF \ln (Ox)/(Red)$

where, E is the potential of the reaction and is measured against the standard Hydrogen electrode.

E_0 is potential (standard reduction potential) in volt when (Ox) = (Red) = 1

F= Faraday constant in heat unit (96,500 coulomb/mol)

R = Universal Gas constant (8.314 J/mol/K); T is the absolute temperature and n is the change in oxidation number

If, E is measured against the standard hydrogen electrode, E is termed as Eh , which is an intensity factor. Therefore,

$$Eh = E_0 + RT/nF \ln (Ox)/(Red)$$

In strongly reducing system the Eh value becomes negative.

$$\text{e.g. } Fe(OH)_{3\ s} + 3H^+_{aq} + e = Fe^{+2}_{aq} + 3H_2O$$

Table 1.2. Oxidised components with redox potential values

Reduction	Redox potential (mV)
$Mn^{+4} \rightarrow Mn^{+2}$	+280 to +220
$Fe^{+3} \rightarrow Fe^{+2}$	+180 to +150
$SO_4^{2-} \rightarrow S^{-2}$	-120 to -180
$CO_2 \rightarrow CH_4$	-200 to -280

pE: It is the measure of the electron activity. Low pE values represent a reducing environment. The pE scale could represent concentration of the standard reducing agent (e-).

$$pE = -\log (e) = Eh/2.303RTF^{-1}$$

$$\text{or, } pE = Eh/0.0591$$

$$or,\ pE_0 = E_0/0.0591 \text{ at } 25^0C$$

if, rH = Reducing Intensity

$$then, rH = -log (PH_2)$$

where, rH = reducing intensity or reducing power of hydrogen

if, PH_2 = Partial pressure of the hydrogen gas (in atm.) in equilibrium with the redox system,

when, PH_2 = 1 atm, rH =0

When the reactant and the product of a reaction are in the gaseous state , then the partial pressure of the gas is considered in formulating the redox potential.

Relationship between; rH , pE and pH

$$\frac{1}{2} rH = pE + pH$$

$$rH = 2 (pE + pH)$$

Crystal structures : The crystal structure of most of the inorganic compounds are governed by the ratio of the number of cations and anions, the ratio of their ionic sizes and properties of polarization of its structural units. i.e., atoms or ions.

1.11 Co-ordination number

In order to attain maximum stability, a compact structures are built up in which every ion is surrounded by as many ions of opposite charge as is possible. The number of nearest neighbouring ions that surrounds one ions of opposite charge is called coordination number and is determined by the radius ratio , i.e. cationic radius (Rx^+) by anionic radius (RA^-), i.e. (Rx^+/RA^-). Thus, co-ordination number may be defined as the number of nearest neighbouring ions that surround one ion of opposite charge in order to attain maximum stable (minimum energy) and compact neutral crystal structure (Table 1.3). The relationship of ions having different valences are observed in respect to radius ratio and coordination number (Table 1.4).

Table 1.3. Relationship between ionic size and coordination number with oxygen for the common cation

Range of Radius Ratio (Rx^+/RA^-)	**Co-ordination number of the central cation**
< 0.155	2
0.155-0.225	3
0.225-0.414	4
0.414-0.732	6
0.732-1.0	8
>1.0	12

$RA^- = 1.32A^0$ (Radius of O^{-2})

Table 1.4. Relationship of ions with radius ratio and coordination number

Ion	Radius(⁰A)	Radius ratio (Rx^+/RA^-)	Observed Co-ordination number
Si^{+4}	0.42	0.30	4
Al^{+3}	0.51	0.36	4,6
Fe^{+2}	0.74	0.53	6
Fe^{+3}	0.64	0.46	6
Ti^{+4}	0.68	0.49	6
Mg^{+2}	0.66	0.47	6
Ca^{+2}	0.99	0.71	6,8
K^+	1.33	0.95	8,12
Na^{+1}	0.97	0.69	6,8
Mn^{+2}	0.80	0.57	6
Li^{+1}	0.68	0.49	6
Cr^{+3}	0.63	0.45	6

Principles Governing the Size Of Ions

1. It is possible when cationic radius is smaller than the radius of the neutral atoms, while anionic radius is greater than the radius of the neutral atom.

   i.e., Anionic radius > Radius of neutral atom > cationic radius
2. In the periodic table, for the elements in the same group, ionic radius increases with the increase of the atomic number (Table 1.5)

Table 1.5. Relation between the atomic number and ionic radius of ion

Group II	Atomic No.	Ionic radius (⁰A)
Be^{+2}	4	0.35
Mg^{+2}	12	0.66
Ca^{+2}	20	0.99
Sr^{+2}	38	1.12
Ba^{+2}	56	1.34

3. For the positive ions of the same electronic configuration , the radius decreases with increase in the charge (Table 1.6).

Table 1.6. Relation between ionic radius and charge of the ions

	Na^{+1}	Mg^{+2}	Al^{+3}	Si^{+4}
Radius (⁰A)	0.97	0.66	0.57	0.42
Atomic Number	11	12	13	14

4. For a multivalent element, higher the positive charge of the ions, smaller is the radius (Table 1.7)

Table 1.7. Relation between ionic radius and ions with multivariances

	Mn^{+}	Mn^{++}	Mn^{+++}	Mn^{++++}
Ionic Radius (^{0}A)	1.18	0.80	0.66	0.60

1.12 Some Aspects of Thermodynamics

The word 'thermo' means heat and 'dynamics' means motion, i.e., heat in motion. So, it is the branch of science which deals with the flow of heat and the relationship between the heat and work.

1.12.1 System

A system may be defined as the specified quantity of matter of an object which belongs to the part of the Universe, chosen as a separate entity from the rest of the Universe, separated by a boundary. All other parts of the Universe outside the boundary of the system are called surroundings.

$$\text{System} + \text{Surroundings} = \text{Universe}$$

Isolated system: A system is said to be an isolated system, when it is incapable of exchanging either energy or matter with the surroundings.

Closed system: A closed system is one in which energy exchange with the surroundings is possible but no mass transfer across the boundary is permitted.

Open system : In open system, the transfer of both mass and energy across the boundary is possible.

1.12.2 Properties of a System

Extensive property

Some of the properties of a system (volume, energy, surface area etc.) are dependent upon the mass of a system, called extensive properties, i.e., if the quantity of material in the system is changed, the value of the property will be changed proportionately.

Intensive property

Some of the properties (temperature, specific heat, refractive index etc.) in the system are independent of the mass of the system called intensive properties.

1.12.3 Functions of a System

State Function: State function is a property of a system , whose change will not depend on the path connecting the two states by which the state of a system is achieved but is determined by the state parameter. e.g., Energy

Non-State Function : Which depends upon the path of a system. e.g. workdone, heat absorbed etc.

The state of a system is expressed by assigning definite values to the thermodynamic properties of a system. If there is any change in a system, one or more of its thermodynamic variables would also change. The Pressure (P), Volume (V) and Temperature (T) determine completely the state of a gaseous system where these state variable are interrelated to each other to define the state of a system .

1.12.4. Heat Capacity

It is defined as the quantity of heat required to raise the temperature of a system by one degree. If the system consists of a single substance, or a solution and its weight is 1 g, then the heat capacity is called specific heat capacity or specific heat. For one mol of a pure substance, the heat capacity of a system is the molar heat capacity or the molar heat of the substance concerned.

Molar Heat Capacity = Molecular weight x Specific heat

1.12.5 Laws of Thermodynamics

First law : The first law of thermodynamics deals with the law of conservation of energy. It states that, although, energy may be converted from one form to another, it can not be created or destroyed. In other sense, whenever a quantity of one kind of energy disappears, an equivalent amount of other kind must be produced.

Exception: Einstein has shown that matter and energy are equivalent and the interrelationship is given by the equation,

$$E=mc^2$$

where, E= energy, m= mass, c = velocity of light

Thus,

1 g of matter can produce energy as ;

$1\text{x}3^2\text{x}(10^{10})^2/4.18\text{x}10^7 = 2.1\text{x}10^{13}$ cals

Equation of 1st law of thermodynamics :

$$Q = W + \Delta E \quad (1)$$

Where, Q = heat adsorbed

W = work done

ΔE = increase of internal energy

If the above change in equation (1) is too small,

$$\delta q = \delta w + dE \quad (2)$$

if the system is under pressure (P) and increased by a small volume dV, then the work done will be PdV and the equation becomes

$\delta q = PdV + dE$ (3)

when the work done by the system is due to volume change (i.e., mechanical work).

Second law of thermodynamics: It is impossible to construct a machine functioning in a cycles which can convert heat completely into equivalent amount of work without producing changes elsewhere.

Third law: It is impossible to attain absolute zero of temperature, though we can approach the absolute zero. Because of this, it is not possible to attain efficiency of the cyclic process to unity.

Zeroth law: If A and B are in thermal equilibrium in a systems and B is in thermal equilibrium with C, then A and C will also be in thermal equilibrium.

Thermodynamic laws are axioms - As there is no proof of thermodynamic laws, nor are there examples known to disprove them, with no exception of any of these laws, they are called 'axioms'.

1.12.6 Chemical Potential of Water

Free energy (G) per mole (n) of a component in a mixture at a given temperature and pressure is called the chemical potential(μ).

$$\mu = \delta G/\delta n \text{, at constant temperature and pressure}$$

- Chemical potential is an intensive property
- Free energy is an extensive property
- Chemical potential of water in ice and that in water at equilibrium at constant T and P must be the same.

If dn moles are transferred from phase I to phase II at constant temperature and pressure, then,

$$dG = [\mu\,(II)\,dn - \mu\,(I)\,dn]$$

But for equilibrium at constant T and P , dG=0

Therefore, $\mu\,(I) = \mu\,(II)$

1.12.6.1 Enthalpy or Heat Content

Sometimes, while dealing a system at constant pressure , a function called heat content or enthalpy (H) is used in place of internal energy.

Equation of Enthalpy;

$H = E + PV$ (1)

Where, P and V are the pressure and volume respectively by the system

Since, E is the characteristics of the system (state function),

'H' is also a state function and the change of heat content (ΔH) would depend only on the initial and final states of the system.

If the final heat content is H_2 and initial H_1,

$$\Delta H = H_2\text{-}H_1 = (E_2 + P_2V_2) - (E_1 + P_1V_1)$$
$$= (E_2\text{-}E_1) + (P_2V_2 - P_1V_1)$$

Therefore, $$\Delta H = \Delta E + P\Delta V$$

When pressure is constant,

$$\Delta H = \Delta E + P(V_2\text{-}V_1)$$
$$= \Delta E + W \text{ (work done)}$$

again, from the 1st law,

$$Q = \Delta E + W \text{ (Heat absorbed)}$$

Hence, $\Delta H = Qp$ (Heat adsorbed at constant pressure)

Heat Capacity

Cv = Heat capacity at constant volume, i.e., heat required to raise the temperature of a substance by one degree Kelvin at constant volume. i.e., change in internal energy per unit increase in temperature (T) at constant volume.

$Cv = (\delta E/\ \delta T)_V$

Cp = Heat capacity at constant pressure, i.e., heat required to raise the temperature of a substance by one degree Kelvin at constant pressure. i.e., change in enthalpy per unit increase in temperature at constant pressure.

$$Cp = (\delta H/\ \delta T)_P$$
$$C_P - C_V = R \text{ (Molar gas constant)}$$

1.12.6.2 Isothermal and Adiabatic Change: The difference between the isothermal and adiabatic change is that, in the former case, the temperature remains constant and the systems exchange heat with its surroundings, while, in the latter case, there must be no heat exchange and so the temperature will be altered.

For the isothermal process,

$$PV = \text{constant}$$

and for the adiabatic process

$$PV^Y = K$$

$$\frac{RT}{V} \cdot V^Y = k$$

$$TV^{Y-1} = \frac{k}{R}$$

$$TV^{Y-1} = constant$$

K and k are constants and $Y = C_p/C_V$

again, $P_1V_1 = P_2V_2$ for isothermal change

i.e., increase in volume is less in adiabatic than in the corresponding isothermal expansion. Adiabatic PV- curve is steeper than isothermal PV- curve (Figure 1.7).

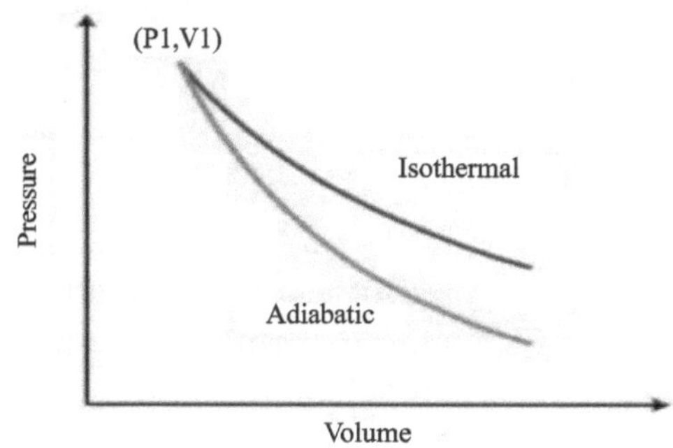

Figure 1.7. Adiabatic and Isothermal PV-curve

Mechanical Equivalent of Heat(J) : This is the ratio of the work done to the corresponding heat change. SI unit of Mechanical equivalent of heat (J) is Joules/calorie (J calorie^{-1}).

$$J = \frac{W}{Q}$$

W = mechanical energy or work
Q = heat generated

Cyclic process: If as a result of series of changes, the systems returns to its initial state, its thermodynamic parameters would also return to their original values. Such a process is called the cyclic process.

Reversible process: A reversible process has been defined as the succession of equilibrium states. A very minute change in one direction or the other of the factors, i.e. temperature or pressure determining equilibrium will cause the system to change in one direction or in opposite direction. A process performed in this manner is said to be thermodynamically reversible.

Irreversible process: An irreversible process is one which occurs suddenly or spontaneously without the restriction of occurring in successive stages of infinitesimal quantities. The system undergoing a change, will not have the tendency to return to the initial state by itself.

Entropy: The entropy difference between the two states of a given system is equal to the summation of the q_{rev} /T terms of all the steps necessary to bring about a reversible (Figure 1.8) transformation between these two states, where, q_{rev} stands for the heat absorbed reversibly and T the corresponding temperature.

$$dS = dq_{rev}/T$$

dS = change in entropy

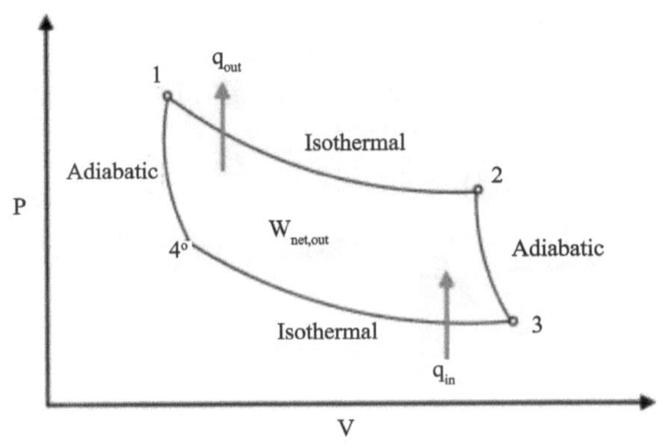

Figure 1.8. Carnot Cycle

In a reversible cycle,

$$\underset{A\to B}{\sum q/T} + \underset{B\to A}{\sum q/T} = 0$$

i.e., there is no change in entropy of a system on completion of a reversible cycle.

Equilibrium state and Steady state : A system is said to be in state of equilibrium when

1. Its composition remain fixed and definite.
2. The temperature at all parts of the system is the same and also identical with that of the surroundings .
3. There is no unbalanced force operative between different parts of the system or between the system and the surroundings. Therefore, a definite temperature, pressure and composition within the system at equilibrium are the criteria to be considered.

In a steady state condition, the overall reaction rate is determined by the slowest of the different steps in a stepwise complex reactions. In most of the cases, the intermediate or unstable species produced in the stages would have a short lived transient existence. At any point of time, the concentration of the intermediates will be quite small and fixed that balances the rate (dC/dt) of formation and disappearance of the intermediates, i.e., $dC_{int}/dt = 0$.

Thus, the intermediate with constant concentration is commonly known as steady state concept.

1.13 Solubility Product

When we take an example of the following,

i.e., dissolution of a sparingly soluble salt (e.g. AgI) in water

$$\mathbf{AgI(s) + H_2O\ (l) \leftrightarrow Ag^+\ (aq) + I^-\ (aq)}$$

Since, this is a dynamic equilibrium, there will be an equilibrium constant (K),associated with it,

$$K = [Ag^+]\ [I^-]/[AgI][H_2O]$$

If the concentration of a solid is 1 mol dm^{-3} and the concentration of water in pure water is 55.6 mol dm^{-3}

Then, Ksp = 55.6 = $[Ag^+]\ [I^-]$, Ksp = solubility product for AgI.

Reactants and Products

$$Zn(s) + H_2SO_4\ (aq) \rightarrow ZnSO_4(aq) + H_2\ (g)$$

Zinc (as solid) reacts with H_2SO_4 in an aqueous solution are reactants where the products are an aqueous solution of the salt $ZnSO_4$ and hydrogen in gaseous form.

1.14 Reaction rate: It is the change in concentration of one of the reactants or products in unit time

$$Zn(s) + H_2SO_4\ (aq) \rightarrow ZnSO_4(aq) + H_2\ (g)$$

If the change in concentration of H_2SO_4 is $\Delta[H_2SO_4]$ in time interval Δt

$$Rate = \Delta[H_2SO_4]/\ \Delta t$$

At the start of the reaction (t=0), the slope of the tangent of the curve is considered as the rate of the reaction.

1.14.1 Factors Affecting the Rate of Reaction

1. Concentration: The greater the concentration of a reactant, the greater the rate of reaction.
2. Pressure: The reaction rate increases with the increase of pressure in the gaseous phase.
3. Temperature: The rate of reaction increases with the increase in temperature.

4. Catalysts : It can boost the reaction up or down without undergoing a permanent change in the process in either phase of the reactions.
5. State of subdivision: The rate of reaction will be enhanced by the fragmentation of the particles for increasing surface area.

1.14.2 Endothermic and Exothermic Reactions

In any chemical reaction, there is change in the energy symbolized as ΔH (heat of reaction).

It is found that (Figure 1.9), the reactants have higher energy level than the products and hence, $\Delta H < 0$ (Exothermic reaction) and energy is released (as heat energy) during the conversion of the reactants into products.

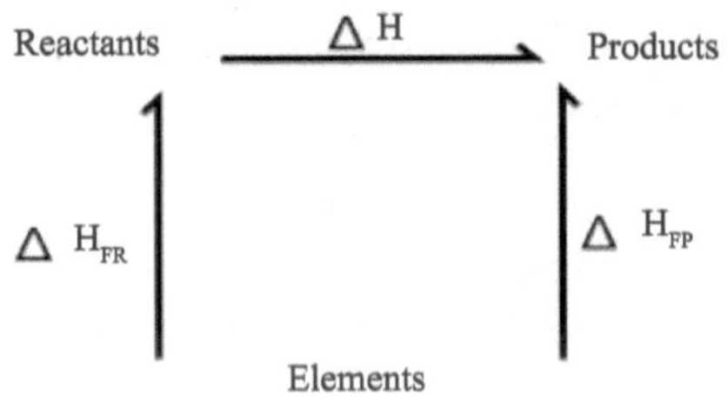

Figure 1.9. Energy involved in the reactants and products of a reaction

On the other hand, when ΔH is positive (i.e., $\Delta H > 0$), the reaction is endothermic and energy is taken up (as heat) during the conversion of reactants into products.

Energy of activation

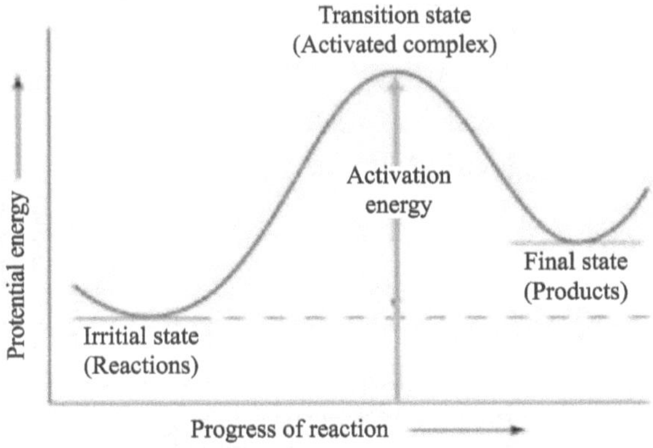

Figure 1.10. Activation energy and progress of reaction

It is observed (Figure 1.10) that, there is a 'hump' in the curve representing the energy barrier (Ea) while there is a change in energy from the reactants to products. The energy of activation (Ea) has role to play in the reaction rate. If the Ea is high, then the reaction rate will be relatively slower compare to the small Ea value. At the peak of the curve there exists a transition phase where the reactants and the products loose their identity and can be converted either. A catalyst increases the rate of reaction by reducing the energy of activation of that reaction.

1.15 Radioactivity

It is natural and spontaneous process by which the unstable atoms of an element can radiate the excess energy in the form of particles or waves. The emissions are collectively called the ionizing radiations. During the decay of a radioactive substance, there is ionization of an element through radiation. This emitted radiation with higher energy when collide with other elements can remove electrons from the atoms in the material.

1.15.1 Alpha (α) Particle

In a radioactive process a particle with two neutrons and two protons is ejected from the nucleus of a radioactive atom. The particles (Table 1.8) is identical to the nucleus of helium atom. As alpha particle contain two protons , they have a positive charge of two.

$$U_{235} - {}^{4}_{2}\alpha^{++} \text{ --------------------} \rightarrow Th_{231}$$

1.15.2 Beta (β) Particle

In a radioactive process when an electron is emitted from the nucleus of a radioactive atom along with an unusual particle called an antineutrino, then β particle is emitted.

${}^{0}_{0}\nu$ (antineutrino)

↑

K_{40} -------------------- → ${}^{0}_{-1}\beta^{-}$

(Parent nucleus)

↓

$\mathbf{Ca_{40}}$

(Daughter nucleus)

The neutrino particle carries away some of the energy from the decay process. The β particles have a single negative charge and weigh only a small fraction of a neutron or proton. Hence, beta particles interact less readily with material than alpha particles.

1.15.3 Gamma – Ray (γ) Radiation

On completion of the decay process, sometimes the nucleous of an atom get excited with the excess energy which is lost by emitting a pulse of electromagnetic radiation called a gamma ray which is identical in nature of light or microwaves with high energy. The gamma ray does not have mass or any charge.

$\uparrow^{0}{}_{-1}\beta^{-1}$

$C_{0}60$ -------------------------- ---------------$\rightarrow Ni_{60} \rightarrow \rightarrow$

(daughter nucleus) (parent nucleus)

Table 1.8. Properties of radiation

Radiation	Type of Radiation	Mass (AMU)	Charge	Shielding material
Alpha	Particle	4	+2	Paper, skin,cloths
Beta	Particle	1/1836	±1	Plastic,glass,light metals
Gamma	Electromagnetic wave	0	0	Dense metal, concrete,earth
Neutrons	Particle	1	0	Water, concrete, polyethylene,oil

1.15.4 Half life and half decomposition of radioactive substances

Half-life: It may be defined as the time required for the quantity of a radioactive material to be reduced to one- half its (Figure 1.11) original value (Half decomposition).

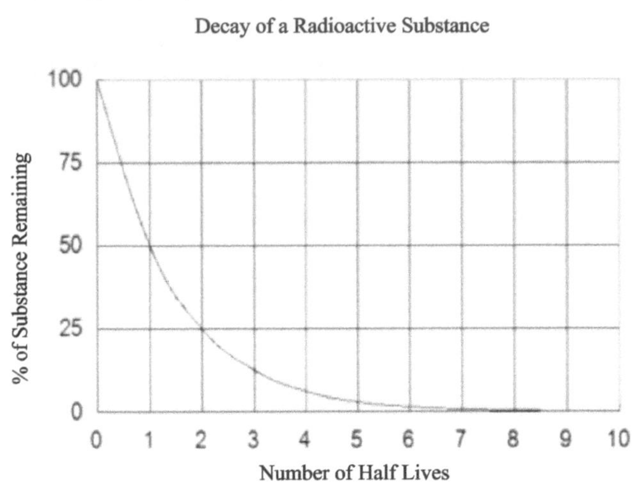

Figure 1.11. Radioactive decay

U_{238} → half life → 4.5 x 10^9 years

C_{11} →half life → 20 minutes

Measures of radioactivity

Activity: It is the number of disintegration or transformations the quantity of material undergoes in a given period of time.

Curie (Ci): 3.7 x 10^{10} dps (disintegration per second)

Becquerel (Bq) : 1 dps

Radiation units

Roentgen – a unit for measuring the amount of gamma or X ray in air (coulombs/kg)

Rad- a unit of measuring absorbed energy from radiation (Gray)

Rem- a unit for measuring biological damage from radiation .

Instruments

1. Gas- filled detector
   a) Ion chamber - measures the large amount of radiation
   b) Geiger - Muller or GM detector - measure very small amount of radiation
2. Sodium Iodide Detector/ Scintillation

   Detector : The material (Sodium Iodide) glows or scintillates when radiation interacts with it. The light produced is reflected through a clear window where it interacts with device called a photomultiplier tube.

   Isotopes : The radioactive isotope is also known as the radioisotope or nuclide having same chemical element with different masses, having unstable nuclei and dissipate excess energy by emitting radiation in the form of α, β and γ rays.

   e.g., on decay (Figure 1.12) of U_{238}, it produces Thorium, Radon, Bismuth

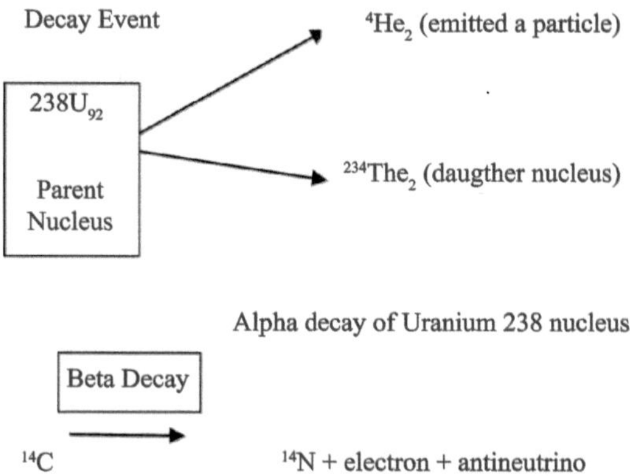

Figure 1.12. Radioactive decay of Uranium and Carbon

Examples of radio isotopes

Carbon: ^{13}C,^{14}C; Phosphorus: ^{33}P; Nitrogen: ^{15}N

1.15.5 Application of Radio Isotopes

Some of the Radio isotopes (^{14}C,^{33}P,^{15}N etc.) are used in the fields of plant physiology, soil chemistry and plant biochemistry as Radiotracer techniques through which small amount of labelled nutrient can be determined in the bulk of nutrients. Besides, the isotope dilution analysis (IDA) is also used in chemical analysis which is based on the mixing (ordilution) of a radioisotope or a separated stable isotopes with its natural isotopes in the sample. In the original mixture the total quantity of the substance can be determined from the fraction of the libelled isotope present in the sample.

If the known amount (m_1) of isotopic tracer with the activity A_1 is mixed with M_1 (weight of the unlabelled compound) and the activity of an aliquot Z_1 from the mixture is A_2

then, the specific activity of the tracer, $S_1 = A_1/m_1$ and the specific activity (S_2) of the mixture (M_1+m_1) will be $S_2 = A_2/Z_1$

If the total radioactivity is unchanged,

Then, $S_1m_1 = S_2 (M_1+m_1)$

$$M_1 = m_1 (S_1/S_2-1) = Z_1 (A_1/A_2)-m_1 \quad (1)$$

Since , $m_1 << M_1$, the equation (1) can be written as ,

$M_1=m_1.S_1/S_2 = Z_1A_1/A_2$

where, A_1/A_2 is the fractional recovery of the isotopic tracer in the weight Z_1 of the sample.

Questions

A. Objective: (1 mark)

1. The salt NaCl is having
   a) Ionic bond b) Covalent bond
   c) Dative bond d) None of the above
2. Sharing of outermost electrons between atom could form
   a) Covalent bond b) Ionic bond
   c) Hydrogen bond d) Dative bond
3. The molecular geometry of F-Be-F is
   a) Linear b) Pyramidal
   c) Cubical d) None of the above
4. Molecular arrangement of methane is
   a) Linear b) Tetrahedral
   c) Pyramidal d) Trigonal planer
5. The molecular weight of H_2SO_4 is
   a) 98 b) 99
   c) 100 d) 96
6. If the radius ratio of a molecule is between 0.225-0.414, the coordination number of the central cation will be
   a) 6 b) 4
   c) 8 d) 10
7. Energy of a system is
   a) Extensive property b) Intensive property
   c) Surface property d) None of the above
8. Work done by the system is
   a) Non-state function b) State function
   c) Stoichiometric function d) None of the above
9. Thermal equilibrium in a system deals with the
   a) 1st law b) 2nd law
   c) 3rd law d) 4th law of thermodynamics.

10. The entropy in nature is
    a) Increasing
    b) Decreasing
    c) No change
    d) None of the above
11. Enthalpy or heat content is equivalent to heat adsorbed at
    a) Constant pressure
    b) Constant temperature
    c) Constant volume
    d) None of the above
12. When free energy change (Δ G) is negative, the reaction is
    a) Spontaneous
    b) Non-spontaneous
    c) At equilibrium
    d) Zero order reaction
13. The half life of U_{328} is
    a) $4.5x10^9$ years
    b) $4.5x10^{10}$ years
    c) $4.5x10^{11}$ years
    d) $4.5x10^8$ years
14. The radio isotope of nitrogen is
    a) 14_N
    b) 15_N
    c) 16_N
    d) 18_N
15. $TV^{\gamma-1}$ in an adiabatic process is
    a) 0
    b) Constant
    c) 1
    d) >1

B. Subjective: **(2 marks)**

1. Outline different types of chemical bonds found in atoms.
2. Define hydrocarbons with examples.
3. What is the correlation among Rh, pE and pH ?
4. What do you mean by coordination number ?
5. What are intensive and extensive properties under thermodynamic principles.
6. What could be the maximum efficiency of a machine ?
7. Differentiate between the steady state and equilibrium state.
8. What is energy of activation in a reactant product equilibration ?
9. What do you mean by half decomposition of a radioactive decay ?
10. What is radio isotopes ?

C. Descriptive: **5 marks**

1. Describe briefly with examples the function of ionic and covalent bonds.
2. What do you mean by molecular weight ? Outline the relation between atomic weight and valency.
3. Describe in brief the oxidation- reduction (Redox) potential in a system.
4. Outline the correlation between the radius ratio of ions with the corresponding coordination number.
5. What are the principles governing the size of ions.
6. Outline different laws of thermodynamics with examples.
7. Why does adiabatic PV-curve steeper than isothermal one ?
8. Briefly describe the factors affecting the rate of a reaction.
9. Outline in brief the endothermic and exothermic reactions.
10. What are the applications of radio isotopes in agriculture.

Suggested Readings

Glasstone, S. (1946) Textbook of Physical Chemistry. Macmillan student editions
Rakshit, P.C. (2001) Physical Chemistry. Publisher, Sarat Book Distributors ISBN
Palit, S.R. (1938) Elementary Physical Chemistry. Book Syndicate Pvt. Ltd., Calcutta.

2

Composition of Soil

Soil is one of the major platforms on which life sustains. Soil, the product of weathering, covers the earth's surface. Soil is the medium of plant growth and thus it requires proper attention for better crop yield. The land is the constituent of soil, vegetation, physico-chemical and biological attributes affecting its usages. The soil may act as a sink or source for the elements and other materials. Hence, it requires to understand the various components to facilitate management options for improving the fertility status of soil.

2.0 Soil

The soil is a natural dynamic body on which plant grows. It is a complex mixtures of minerals, water, air, organic matter and microorganisms. Besides, there exist an unconsolidated mineral or organic material on the surface of the earth and thus becomes a natural medium for the growth and nutrition of the plants.

Jenny (1941) considered soil as a naturally occurring body that has been evolved owing to combined influence of climate and organisms acting on parent materials, as conditioned by relief over a period of time.

Joffe (1949) considered soil as a natural body of mineral and organic constituents, differentiated into horizons of variable depth, which differs from the materials below in morphology, physical makeup, chemical properties and composition and biological characteristics.

2.1 Branches of Soil Science

There are different branches of soil science, each depicting a particular aspects of this natural resource governing the production potential of the crops. The branches are:

i) **Soil Chemistry:** It deals with the chemical constitution and properties involved while developing soil from the rocks. The chemical reactions involved while transport of nutrients at its plant available pool are being dealt in this section.

ii) **Soil Fertility:** It deals with the supplying capacity of the essential plant nutrients (micro and macro nutrients) for growth and nutrition of the plants. Soil fertility is related with the production potential of the soil for a particular field.

iii) **Soil Physics:** It concerns with the physical properties of soils and regulations of physical processes (e.g., state of a system and transport of matter) within and through the soil.

iv) **Soil Biology:** It refers to the soil-born organisms and their lifecycles and activities. It deals with the roles played by the organisms (bacteria, fungi, actinomycetes, insects, nematodes etc.) in transforming nutrients and carbon stock and thus affecting the fertility status of soils.

v) **Soil Mineralogy:** This branch deals with the primary and secondary minerals of the soils and their genesis and classification while understanding the contribution of mineral ore to a particular zone.

vi) **Soil Genesis and Classification (Pedology):** Soil genesis deals with the weathering of rocks and minerals and factors and processes of soil formation. The soil classification is the systematic and recognised rearrangements of soils into different categories based on the soil characteristics.

vii) **Soil Survey:** The morphological assessment of the soils in the field and its analysis in the laboratory, their description, classification, mapping and interpretation based on their suitability under management systems is described in soil surveys. The soil survey provides the inventory of soil resources and is important for any country.

viii) **Soil Salinity:** It concerns with the excess soluble salts (carbonate, bicarbonate, sulphates, chlorides etc.) present in the soils, their reclamation and management to bring this problem soils under cultivation.

ix) **Soil Conservation:** It deals with the conservation of natural resources including soil from the physical loss by erosion (by wind and water) or chemical deterioration. Hence, management and land use method are important to safeguard the soil against deterioration by natural or by induced factors.

x) **Soil Microbiology:** It deals with the microbial population in soil, their classification and roles with respect to soil fertility and land reclamation. It concerns with the nutrition of plant as well as understanding the diseases caused by any soil-born plant pathogens.

2.2 Components of Soil

Soil in general having the three phase systems (solid, liquid and gaseous) are composed of minerals, organic matter, water and air (Figure 2.1). A typical soil consists of approximately 45% mineral, 5% organic matter, 20-30% water, and 20-30% air. The proportion of air and water will vary depending upon the moisture status of the soil.

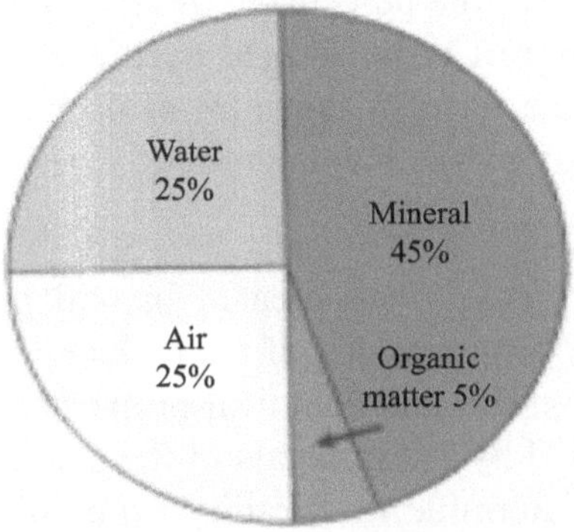

Figure 2.1. Volume composition of soil

2.2.1 Soil Water

Soil water is held within the soil pore spaces with varying degrees of tension or suction which restricts the movement of water through the soil. Soil water dissolving the soluble constituents can form the soil solution, being the medium for supplying the soil –nutrients to plants. Soil water is essential for the survival of the plants. There are three main types of soil water - gravitational water, capillary water and hygroscopic water based on the function of the water in the soil.

a) **Gravitational water**: Gravitational water is the free water moving through soil by the action of gravity. It is largely found in the macropores spaces of soil and very small amount of it is available to plants for rapidly draining out beyond the rhizosphere which percolates down into the water table . But, in a compact soil , the gravitational water is relatively available to the plants. The water is held by the soil with the matric potential between zero and -0.1 (for coarse textured) or -0.33 (for fine textured) bar.

b) **Capillary water**: Capillary water is held in the micropore spaces of the soil. The surface tension properties (cohesion and adhesion) of the soil micropores are stronger than the force of gravity which causes holding of this type of water. However, when the soils dry out, the capillary water goes down the soil under the force of gravity. Capillary water close to the region of root rhizosphere is available to plants. The matric potential of this part of water varies from soil types but is generally in the range of -0.1 or -0.33 bar to -31 bar.

c) **Hygroscopic water:** Hygroscopic water forms as a very thin film surrounding the soil particles and is generally not available to the plant. The adhesive force is so strong that very little of it can be taken up by plant roots. In certain clayey types of soils , the percentage of this type of water is relatively higher. The fraction of water is held by the soil particles by suction beyond 31 bar in the form of thin film of 4 or 5 molecules. This water can be utilized only by the few microbes.

2.2.2. Soil Mineral

During the formation of the earth, different rocks were created in which there were several types of minerals. With the disintegration of the rocks and minerals, soils are formed having different physical and chemical properties of these rocks. The clay fractions are dominated by the minerals of the soils. The characterization of soil minerals could determine the fertility status of soil. The soil environment contains the primary and secondary minerals. Due to the development of the electrical charge , the soil mineral can adsorb plant nutrients on its surface. With the change in chemical conditions, nutrients and other compounds can be released back into the soil solution which are available to the plants and soil organisms. Some of the most common *minerals* found in *soil* are rich source of elements like Iron, Potassium, Magnesium, Calcium, Sulphur etc. The persisting primary minerals (quartz, feldspar) could change in composition on exudation of molten magma in different soil textures. On the other hand, the secondary minerals (silicate clays, oxides of iron and aluminium) are formed by breakdown and weathering of less resistant minerals. The secondary minerals are mostly dominant in clay fraction (<0.002 mm diameter).

2.2.3 Soil Air

The volume of the soil not occupied by soil solids is known as 'Pore Space'. The pore space is usually occupied by air and water. The air filled pores spaces are filled up with the gaseous phase of the soil system. It is a source of oxygen for aerobic microorganisms and plant roots, where, air is available as free, absorbed and in dissolved phase. Oxygen is essential for all biological reactions occurring in soil. The soil air contains relatively higher volume of carbon dioxide than oxygen compared to the atmospheric air. Besides, it also contains, nitrogen and water vapour. The level of carbon-di-oxide in soil increases on respiration by soil fauna and flora and or addition of organic substances. The liberated oxygen in atmosphere in turn is taken up by the animals for instance. The soil fauna and flora consume oxygen and gives out carbon dioxide. Organic matter addition and cropping increases the level of Carbon dioxide in the soil. Carbon dioxide in soil air can move from the soil pores into atmosphere and vice-versa. The nitrogen content in soil air is almost the same as in atmosphere (Table 2.1). There is a continuous gaseous exchange between the atmospheric air and soil air . The circulation of air in the soil mass is known as 'soil aeration'. The movement of soil air is governed by far with temperature, diffusion, wind and water.

Table 2.1. Composition of soil air

Component	Soil air (%)	Atmosphere (%)
N_2	79.2	79.0
O_2	20.6	20.9
CO_2	0.25	0.03

Source: Russel and Appleyard (1915)

2.2.4 Soil Organic Matter

Soil organic matter (SOM) is one of the major components of soil, consisting of plant and animal residues at various stages of decomposition of microbial cells and tissues, synthesised products of soil microbes (Figure 2.2) . A fertile and healthy soil is the basis for healthy plants, animals, and humans for which soil organic matter build-up is needed for developing healthy and productive soils. The organic matter in soil can take a pivotal role for improving soil health and resilience. The build-up of soil carbon stock and its impact on global carbon budget can reflect the role to be played by the soil organic matter. The SOM can promote the soil processes and crop growth as a critical component in the systems. Besides, the products from advanced decomposition of organic residues and or re-synthesised products of microorganisms called 'humus' has high surface area per unit mass which could affect the ion exchange processes in the soil systems. The SOM is considered as the store house of micro and macro elements as well as some secondary and pollutant elements. Soil organic matter basically have two groups; a) original tissues and their partially decomposed products and b) humus. The accumulation of organic matter under wet anaerobic condition is considered as peat.

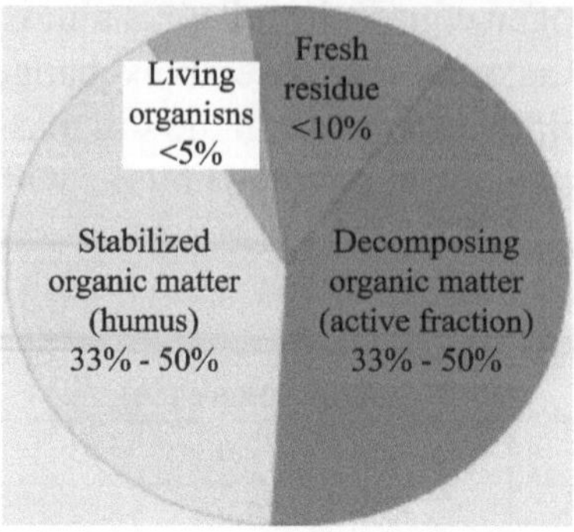

Figure 2.2. Sources of organic matter in soil

The soil structure, moisture holding capacity, diversity and activity of soil organisms and nutrient availability are influenced by the SOM. It also influences the effects of chemical amendments, fertilizers, pesticides and herbicides in soil.

2.3 Soil Profile

The soil is formed with different layers arranged during the formation of soil. These layers are called *horizons*. The sequence of layers is called *soil profile*. The layers of soil can easily be identified by their colour and size of particles. The layers of the soil are mainly composed of topsoil, subsoil and the parent rock. The soil profile is defined as a vertical section of the soil which is exposed by a soil pit. A soil pit is a hole that is dug from the surface of the soil upto the underlying bedrock.

2.3.1 Components of the Soil Profile

A **soil horizon** making the separate layer of soil, runs parallel to the soil surface and has different properties and characteristics than the adjacent layers above and below the soil layers. The soil profile extends from the soil surface to the parent rock material. The unconsolidated debris is called the regolith, which includes all of the weathered material within the profile. The regolith has two components: the **solum** and the **saprolite**. The solum includes the upper horizons with the most weathered portion of the profile. The saprolite is the least weathered portion that lies directly above the solid, consolidated bedrock but beneath the regolith. In a profile, there are succession of horizons which are differentiated from one another and genetically interrelated in association with the parent materials.

Master Horizons

There are **five master horizons** in the soil profile. The master horizon may not contain all five horizons in all the soil profiles which could differ from one location to another. The five master horizons are represented by the letters: O, A, E, B, and C (Figure 2.3).

O: The O horizon is a surface horizon that is comprised of organic substances at various phases of decomposition. It is most prominent in forest based ecosystem , where there is accumulation of debris fallen from trees.

A: The A horizon is a surface horizon that largely consists of minerals (sand, silt, and clay) and with appreciable amounts of organic matter. This horizon is predominantly the surface layer of many soils in grasslands and agricultural lands. The organic matter enriched A horizon present at the soil surface is known as top soil or surface soil. The upper 15 cm depth of soil which is considered as the plough zone areas/plough layer is the region of plant root's proliferation. The compact soil layers underlying the surface soil are referred to as subsoil. The water and some nutrients which are required by the plants are stored in the sub-soil. Subsoil horizons are light in colour due to lack of organic matter and aeration status is also decreased with the depth of soil.

E: The E horizon is a subsurface horizon that has been heavily leached. The horizon is typically light in color and is generally found beneath the O horizon.

B: The B horizon is a subsurface horizon that has accumulated materials from the layer(s) above. It is a site of deposition of certain minerals that have leached down from the layer(s) above.

C: The C horizon is a subsurface horizon and is the least weathered horizon. It is unconsolidated with loose parent materials.

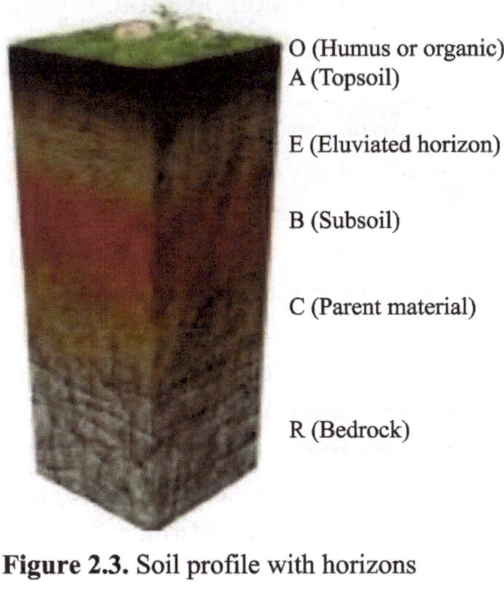

Figure 2.3. Soil profile with horizons

2.4 Rocks and Minerals

A **rock** is an aggregate of one or more minerals, or a body of undifferentiated mineral matter. Common rocks include granite, basalt, limestone, sandstone etc.

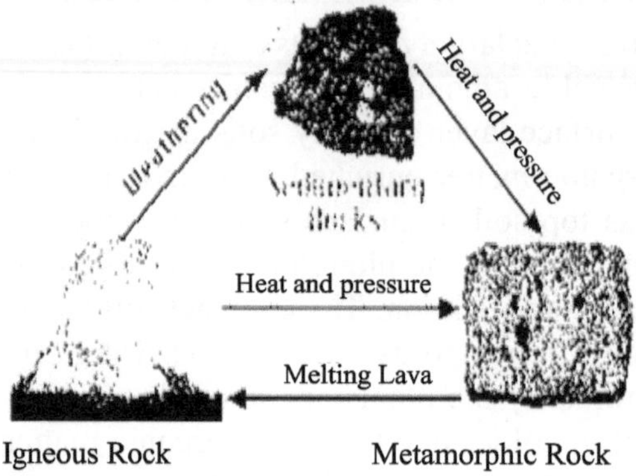

Figure 2.4. Rock cycle

There is a continuous change of the earth's rock with the progress of time as is evidenced from the rock cycle (Figure 2.4) when a certain type of rock gets exposed to weathering processes, it breaks apart over time and is transformed into other type of rock (e.g. molten magma to igneous rock). Subsequently, there may be transformation of this igneous rock into sedimentary rock (conglomerate) or metamorphic rock.

A **mineral** is a naturally occurring inorganic element or compound having an orderly internal structure and characteristic chemical composition, crystal form, and physical properties. Common minerals include quartz, feldspar, mica, amphibole, olivine, calcite, corrundum etc.

A mineral is a solid material, made up of one substance, that occurs naturally on Earth. Most of the common minerals are made of crystals which is a solid formed by a repeating, three-dimensional pattern of atoms, ions, or molecules and having fixed distances between the different parts. The mineral matter consists of portion of the weathered materials like aluminosilicates, free silica etc. and partially un-weathered particles of original rocks and minerals (Quartz, mica etc.). Besides, there may be oxides of Al, Fe, Mn etc.as well as insoluble salts of carbonates. The soluble parts are composed of soluble salts like alkali and alkaline earth metals. Some of the minerals are there which lack the crystalline pattern or definite structure (e.g. allophone, imogolite) are called amorphous minerals, found from the volcanic ash deposition.

2.5. Elemental Composition of the earth's crust :

The Earth's crust is a thin layer on the outer periphery of the Earth, accounting for less than 1% of Earth's volume. It is a part of the lithosphere: a division of Earth's layers that includes the crust and the upper part of the mantle. The tectonic plates developed from the lithosphere can move, allowing heat to escape from the interior of the Earth into the available space. According to Frederick and Edward (2000), earth's crust is made up of several elements (Figure 2.5): oxygen (46.6 percent by weight); silicon (27.7 percent); aluminum (8.1 percent); iron (5 percent); calcium (3.6 percent); sodium (2.8 percent), potassium (2.6 percent) and magnesium (2.1 percent). From the composition there is evidence of the dominance of silicon, aluminium and oxygen (Figure 2.5) over the earth crust forming the silicates. There is formation of silicon tetrahedral and aluminium octahedral structures which are the building blocks of silicate minerals on earth.

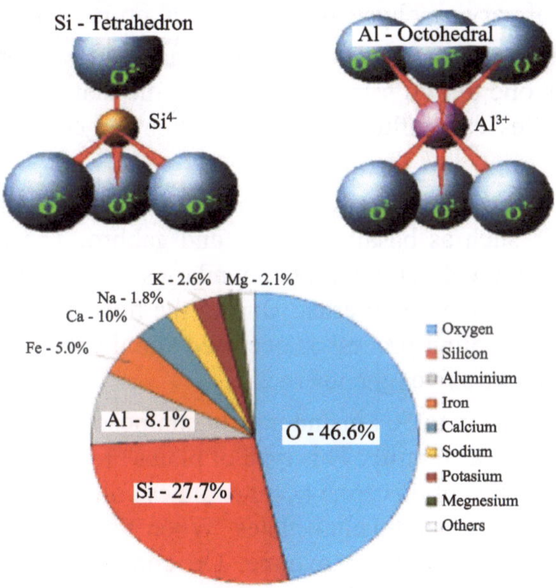

Figure 2.5. Mineral composition of the earth's crust

Abundance of Si and O over the earth's crust

Since O and Si are by far the most common elements on the earth crust (47% and 28%) , the majority of the minerals occurring in rocks are compounds of Si and O – the silicates. Besides, Si is in the group IV of the periodic table and each of its atoms has 4 electrons in the outer shell and requires more electron to complete an inert gas structure Argon. On the other hand, oxygen atoms (Group VI) has 6 electrons in their outer shell and requires further 2 electrons to give the electronic structure of Neon. The basic unit of all silicates is the Si-O in tetrahedron in which a silicon atom is bounded by single covalent bonds to four oxygen atoms arranged at the corner of a tetrahedron.

$$Si + 4\,O = SiO_4$$

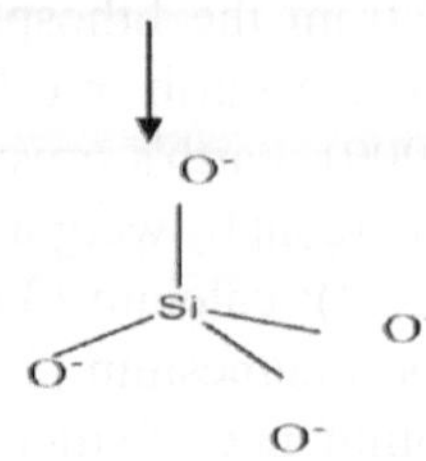

In this way, each Si atom obtains a share in a stable octet of electrons but each oxygen is still one electron short of the neon electron structure . This deficit of electrons may be made in a number of ways leading to a variety of naturally occurring silicates compounds of varying physical and chemical properties.

There is a good number of rigid pieces on earth's surface with major and minor tectonic plates. These plates are one of two primary types: oceanic plates and continental plates. The two plates can influence the fundamental geologic processes of our planet.

1. **Oceanic:** Thickness is 5 km to 10 km and composed primarily of denser, more mafic rocks, such as basalt, diabase, and gabbro. These type of plates are formed by the divergent plate boundaries through the upwelling magma creating new oceanic crust. During the flow of lava from the volcanic ridges, a cooling process is occurred simultaneously leading to the formation of extrusive igneous rock.
2. **Continental:** Thickness is 30 km to 50 km and mostly composed of less dense, more felsic rocks, such as granite. This type of plates are formed by convergent plate boundaries. In this zone oceanic plates collide with and plunge underneath with the continental plates. When the oceanic plates are subducted, they melt to form magma. On cooling of the magma, intrusive igneous rock is formed and new continental crust is developed.

Questions

A. Objective

1. In which of the following soils, water infiltration is more rapid ?

a) Sandy soils b) Loamy soils

c) Clay soils d) All

2. Which soil structure is most desirable for agriculture?

a) Platy b) Blocky

c) Crumbs d) Granular

3. Element essential for deflocculation of soil particles is

a) Potassium b) Calcium

c) Magnesium d) Sodium

4. Minerals that crystallize from cooling magma are

a) Primary minerals b) Secondary minerals

c) Tertiary minerals d) Rocks

5. If pH reading is zero for any suspension, what is its H^+ ions concentration in mol/L

a) 1 b) 2

c) 3 d) 4

6. Most dominant element in Earth crust is

a) $O_{2;}$ b) Ca^{++}

c) Al^{3+} d) Si^{4+}

7. Which of the following is a ferromagnesian mineral?

a) Anorthite b) Quartz

c) Olivine d) Albite

8. Which of the following is a non-silicate mineral?

a) Kaolinite b) Calcite

c) Amphibole d) Vermiculite

9. Example of metamorphic rock is

a) Granite b) Rhyolite

c) Shale d) Slate

10. Which of the following mineral is most easily weatherable

a) Feldspar b) Quartz

c) Gypsum d) Zircon

11. Which of the following is an active soil forming process?

a) Climate b) Parent material

c) Time d) Relief/ topography

12. Deposition carried by ice-glacier is called
    a) Colluvium b) Moraine
    c) Alluvium d) Loess
13. Among the below four stages, soil of which stage is most mature?
    a) Senile b) Initial
    c) Virile d) Juvenile
14. Which of the following is a process of removal of soil constituents from upper horizon in suspension/ solution form?
    a) Humification b) Illuviation
    c) Calcification d) Lessivage
15. Soil forming process leads to leaching of bases is called
    a) Podzolization b) Calcification
    c) Humification d) Pedoturbation

B. Subjective

1. Define soil and outline the different branches of soil science.
2. What are the composition of soil ?
3. What is the importance of soil organic matter in agriculture ?
4. What is soil profile ?
5. Differentiate between rocks and minerals.
6. What are the dominant ions on earth crust ?
7. What is continental drift ?
8. What is the correlation between the depth of earth crust and sub-soil temperature?
9. Why there is abundance of Si and O on earth crust ?
10. What are the building blocks of silicate layers ?

C. Descriptive

1. What is soil water ? Briefly outline the different types of soil water.
2. What are the composition of soil air and atmospheric air ? what is the importance of soil air ?
3. Briefly describe the components of soil profile.
4. Describe the composition of earth crust.
5. Why there is a need on natural resource conservation ?

6. How soil fertility can be improved ?
7. What are the soil moisture constants ?
8. How does soil develop from rocks and minerals ?
9. Describe the sources of soil organic matter.
10. Why soil is considered as a heterogenous substances.

Suggested Readings

Frederick K. Lutgens and Edward J. Tarbuck (2000). Essentials of Geology. 7th Ed., Prentice Hall.

Joffe, J.S. (1949). Pedology. New Brunswick, New Jersey, Pedology Publication.

Dey, N.K. and Sarkar, H.K. (1993). Soil Geography. Shribhumi Publishing Company, Kolkata of variation. J. Agr. Sci. 7:1-48

Russel, E.J. and Appleyard, A. (1915). The atmosphere of the soil, its composition and causes Simonson, R.W. (1968). Concept of Soil. Adv. Agron. 20:1-47.

3

Elements in Plant Nutrition

The plant can absorb more than sixty elements as revealed from the plant tissue analysis, still a few elements is considered essential for the growth and nutrition of the plants. The deficiency of any one of the essential elements can stop completing the life cycle of the plants which is reflected on the yield of the crop. The growth of the plant which is expressed in terms of dry weight, length ,diameter etc. are interlinked with the plant available nutrients of soils.

3.1 Essential Nutrients

The essential elements are those elements which are required for growth and nutrition of the plants. There are 17 elements known to be essential nutrients for plants. The essential elements are classified as micro and macronutrients depending upon the requirement of the crop. Among the naturally occurring elements, eighteen elements are considered essential to plants. These are: C, H, O, N, P, K, Ca, Mg, S, Fe, Mn, Zn, Cu, B, Mo, Cl, Ni, Co. The elements which are found in different sources are required to be in available forms (Table 3.1) for growth and nutrition of the plants.

Table 3.1. Plant available forms of elements

Elements	Average concentration in plants (dry weight basis)	Plant available forms
Hydrogen (H)	6.0%	H_2O
Oxygen (O)	45.0%	H_2O and O_2
Carbon (C)	45.0%	CO_2
Nitrogen (N)	1.5%	NO_3^- and NH_4^+
Phosphorus (P)	0.2%	$H_2PO_4^-$ and HPO_4^{2-}
Potassium (K)	1.0%	K^+
Calcium (Ca)	0.5%	Ca^{+2}
Magnesium (Mg)	0.2%	Mg^{+2}
Sulphur (S)	0.2%	SO_4^{2-}
Iron (Fe)	100 ppm	Fe^{+2}
Manganese (Mn)	20 ppm	Mn^{+2}
Zinc (Zn)	20 ppm	Zn^{+2}

Copper (Cu)	6 ppm	Cu^{+2}
Boron (B)	20 ppm	$H_2BO_3^-$, HBO_3^{2-} and BO_3^{3-}
Molybdenum (Mo)	0.1 ppm	MoO_4^{2-}
Chlorine (Cl)	100 ppm	Cl^-
Nickel (Ni)	0.01 ppm	Ni^{2+}
Cobalt (Co)	0.01 ppm	Co^{2+}

Source: Havlin *et al.* (2007) and Brady and Weil (2007)

3.2 Functional Nutrients

There are some elements (Rb, Ca, Stronsium etc.) which can play a specific role in plant metabolic processes, although, their action may not be specific, i.e., the requirement can be met by another element.

3.3 Classification of Essential Nutrients

Essential nutrients are classified into two major groups on the basis of relative utilization or absorption by the plants and also based on their biochemical behavior and physiological functions.

3.3.1 Based on Relative Utilization or Absorption by the Plants

A. Macro Nutrients

B. Micro nutrients

Again, Macronutrients are classified into two types :

1. **Primary Nutrient:** Nitrogen, Phosphorus and Potassium. These three elements are also called as fertilizer elements.
2. **Secondary Nutrients:** Calcium, Magnesium and Sulphur.

3.3.2 Classification of Nutrients Based on their Biochemical Behaviour and Physiological Functions

1. Group I: It includes C,H,O and Ca.

The C,H,O and Ca elements are required for basic structural constituent of plants (carbohydrates,proteins and fats) and organic matter. These elements are not only involved in enzymes processes but also provide energy for growth and development by oxidative breakdown.

2. Group-II: It includes N,P and S.

These elements are used as structural elements for active and living tissues. These elements being components of metabolically active compounds like amino acids, proteins, enzymes and non-proteinaceous compounds are involved in energy storage (ATP and ADP) and transfer (Phosphate esters) processes.

3. Group-III: It includes K, Ca and Mg.

These elements are regulators and carriers for plant metabolism involved in synthesis and translocation of carbohydrates, balancing ionic charges and induce enzyme activation.

4. Group-IV: It includes Fe, Mn, Zn, Cu, B, Mo and Cl.

These elements are used as catalysts and activators and are involved in oxidation-reduction reactions, chlorophyll synthesis and also exists in organic combinations.

3.3.3 Macronutrients

These are nutrients utilized by the plants in relatively large amounts (quantity) for their growth and nutrition of the crops.

i.e., C, H. O. N, P, K, Ca, Mg and S .

Primary nutrients: These nutrients are required for the plants relatively in large quantities for their nutrition and growth. These nutrients are available from the fertilizers other than the soil, i.e., N, P and K.

Secondary nutrients

The elements which are required by plants in moderate amounts are supplied through fertilizers and other amendments; i.e., Ca, Mg and S. When SSP is applied as a fertilizer for P, it supplies Ca and S. When Dolomite is applied as a liming material there is supply of Ca and Mg. Ammonium Sulphate will supply S other than N.

3.3.4 Micronutrients: The nutrients which are required by plants in smaller quantities for their growth and nutrition are called micronutrients which are equally important and essential to plants as macronutrients. They are also called as trace elements.

These include: Fe, Mn, Zn, Cu, B, Mo, Cl and Ni. As the requirement of these element is very low, hence, judicious application is needed to avoid toxicity or deficiency of a particular element.

Beneficial elements

These elements can stimulate the growth of the plants and have beneficial effects at a very low concentration. They are also known as 'potential micro-nutrients'. These elements can affect the uptake, translocation and utilization of other essential elements, stimulates the essential metabolite by activating enzymatic system/action and also counteract the toxic effects of some other elements or anti metabolites ; e.g., Silicon (Si) for rice, Sodium (Na), Aluminum (Al), Cobalt (Co), Selenium (Se), Iodine (I), Gallium (Ga) and Vanadium (Va).

3.3.5 Tracer Elements

Trace elements are micronutrients which are required in minute quantity for growth and nutrition of plants. The tracer elements are radio-isotopes which are used in plants for tracing path of a metabolic reaction. A *radioactive tracer* or radiotracer is a chemical compound in which one or more atoms have been replaced by a radionuclide which can be used to explore the mechanism of chemical reactions by *identifying* the path from reactants to products.

Criteria of essential elements : The concept was coined by Arnon and Stout (1939) and they primarily considered 16 elements essential for plant nutrition. For an element to be regarded as an essential nutrient, it must satisfy the following criteria:

1. A deficiency of an essential nutrient element makes it impossible for the plant to complete the vegetative or reproductive stage of its life cycle.
2. The deficiency of an element is very specific to the element in question which can be corrected /prevented only by supplying that particular element.
3. The element must directly be involved in the nutrition and metabolism of the plant and have a direct influence on plant apart from its possible effects in correcting some micro-biological or chemical conditions of the soil or other culture medium.

3.4 Nutrient Levels in Plant

It is pertinent to mention that, the sustenance of a plant depends on the sources of nutrients in available forms and also on the conditions of the specific zones for the available pools of nutrients. The critical level of nutrient will demarcate the deficient or sufficient zone of that particular nutrient (Figure 3.1).

1. Deficiency zone

The absence of a particular element may cause morphological changes in plants, which are indicative of deficiencies of element and are called deficiency symptoms. The deficiency symptoms of an element can be corrected by supplying that particular mineral nutrient to the plant. The parts of the plants that show the deficiency symptoms also depend on the mobility of the element in the plant. For elements that are actively mobilised within the plants and exported to young developing tissues, the deficiency symptoms tend to appear first in the older tissues. The deficiency symptoms tend to appear first in the young tissues when the elements are relatively immobile and are not transported from the mature organs. Hence, the transport from the sources and mobilization of the nutrients have significant effect in production maximization of the crops for a particular Agro-ecological region.

Critical level

The critical level is defined as the nutrient concentration range in the plant sample below which the yield of crop is significantly reduced. The range may vary from crop to crop and on available nutrient status in soil.

2. Sufficiency zone

Due to the adequate amount of nutrients in soil, there is possibility of having nutrient absorption by the plants at its maximum level. The nutrient concentration range in which added nutrient will not increase yield but can increase nutrient concentration in the vegetative or economic part of the plants. The term 'luxury consumption' is often used to describe nutrient absorption by the plant without influencing the yield of crop.

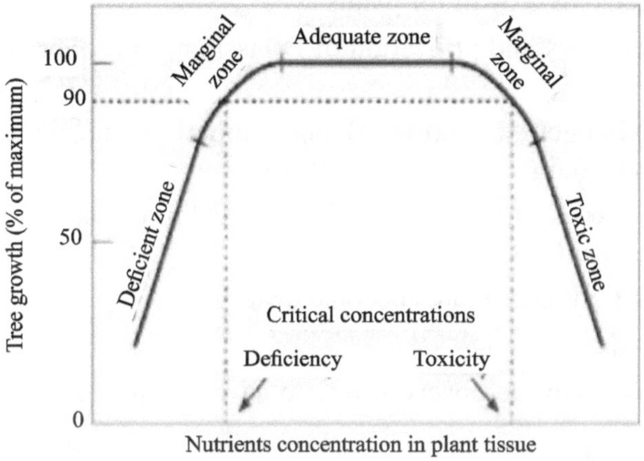

Figure 3.1. Zone-wise nutrient concentration in plants

3. Toxic zone

As is evidenced from the research work that, the gap between the deficiency and sufficiency level of available micronutrients is very narrow, hence, a little bit higher and lower dose of a specific nutrients to the plants may cause toxicity or deficiency symptoms. The gradual accumulation of an element in excess may cause toxicity to the plants causing reduction in dry weight of tissues or may inhibit the uptake of another element. This in tern may reduce the yield of the crops.

3.5 Elements in Plant Nutrition

The primary, secondary and micronutrients are required for the growth and nutrition of the plants. The functions and deficiency symptoms of the specific element is appended below:

3.5.1 Nitrogen (N_2)

Nitrogen is one of the principal growth promoting nutrient elements. Green plants are influenced by the deficiency of nitrogen than by any other element. It is absorbed by plants in the ionic form of NO_3^- by most of the plants and some plants require NH_4^+ form (rice). During the foliar application, NH_2 (amide from) is also absorbed. Nitrogen is playing vital role in plant metabolism in addition to the following important functions:

Functions of Nitrogen in plants

1. Nitrogen is an essential constituent of chlorophyll.
2. It is essential constituent of proteins, nucleotides phosphatides, alkaloids, enzymes, hormones, vitamins etc.
3. It promotes better vegetative growth for plants.
4. It can influence the activities and utilization of Potassium, Phosphorus and other specific elements.

Deficiency symptoms of Nitrogen in plants: Being a mobile element in plants deficiency of nitrogen is mainly observed in older leaves.

1. Yellowing of older leaves due to inhibition of chloroplasts and chlorophyll synthesis causing "Chlorosis" of leaves.
2. Growth of the plants are stunted. Plants become dwarfed.
3. Yield of crop is reduced.
4. Necrosis of plant leaves may be observed due to severe deficiency of nitrogen.

3.5.2 Phosphorus (P)

Phosphorous is a constituent of cell components [phytins, phosphoproteins, phospholipids, nucleic acids (DNA, RNA), co-enzymes, NAD, NADP, ATP] and for the other high energy compounds. Besides, P is a structural component of cell membrane, chloroplasts, mitochondria and meristematic tissues. The mobility of P is relatively less compared to N in plants.

Functions

1. Phosphorus is associated with energy storage and transport.
2. The cell division and development of meristematic tissue is governed by P.
3. Root development and root growth are governed by the availability of P.
4. Phosphorus is required for the formation of starch, proteins and nucleic acids.

5. The photosynthesis, nitrogen-metabolism, carbohydrate metabolism, glycolysis, respiration and fatty acid synthesis are also regulated by P.
6. The seed formation and ripening of fruits are controlled by P.

Deficiency symptoms

1. Growth rate is slow and plants are stunted due to its effects on cell division.
2. Defoliation starts from the older leaves ,leaves' sizes are small. and premature leaf fall occurs.
3. Purplish discoloration of foliage is observed.
4. There may be delay in flowering and ripening of fruits.

3.5.3 Potassium (K)

Potassium is one of the essential elements for fruit crops and field crops. The available form in plants is K^+ ion and it is mobile in nature. It primarily occurs as soluble inorganic salts and occasionally as salts of organic acids and is found in the cytoplasm, meristematic regions, cell sap of plants.

Functions

1. The turger pressure of cell is controlled by K^+ (Osmoregulation)
2. The pH- stabilization, enzyme activation, protein synthesis, stomatal movement, cell extension and photosynthesis are governed by potassium ions.
3. It improves water use efficiency (WUE) of plants under the adverse environmental situation.
4. The ATP synthesis and use efficiency of N are regulated by K^+.
5. Potassium is essential in potato, sweet potato, banana, turnip for formation of starch and sugar.

Deficiency Symptoms

1. Lodging of crops is observed due to the weakening of stem.
2. Burning appearance of leaf margins and leaf tip.
3. Keeping quality of fruits and vegetables are hampered.
4. Marginal necrosis along with the burning of leaf tips.
5. Stunted growth of plants and shortening of internodes are observed.
6. Water use efficiency by the plant is affected.

3.5.4 Calcium (Ca)

The mobility of Ca^{+2} is negligible in plants and exists as deposits of calcium oxalate, calcium pectate in the middle lamella of cell wall and as $CaCO_3$ and $Ca_3(PO_4)_2$ in cell- vacuoles. The requirement of calcium is somewhat less in plant compared to other essential elements.

Functions

1. Calcium is a constituent of the cell wall.
2. The cell divisions and chromosome stability, cell wall construction, cell elongation of the shoot and root are regulated by Ca^{+2}.
3. The Ca^{+2} can impart resistance against diseases by formation of Ca-pectate in the cell wall.
4. Can influence enzymatic activities indirectly.

Deficiency symptoms

As the mobility of Ca^{+2} is slow, hence, deficiency is observed in young leaves and growing tips of plants.

Leaves become small, distorted, crinkled and malformed and root growth of plant is impaired.

The cell wall structure is destroyed affecting nuclear and cell division.

3.5.5 Magnesium (Mg)

Magnesium(Mg) is one of the important constituent of chlorophy ll molecule located at its centre. It is a mobile element and plant can absorb as Mg^{2+} ion. Excess of Mg absorption may lead to browning of roots and growth of plant become restricted.

Functions

1. Magnesium is essential for photosynthesis of plants.
2. The cellular pH, cation-anion balance and turgur pressure of the cells is regulated by Mg.
3. Mg^{+2} is required for protein synthesis.
4. The stabilization of cell membranes is influenced by Mg^{+2}.

Deficiency symptoms

1. Interveinal chlorosis at the lower leaves may be observed.
2. Leaves become small and brittle even at its maturity.
3. Premature dropping of leaves may cause loss of fruit crops.
4. Nitrate reduction is inhibited.

3.5.6 Sulphur (S)

Plant absorbs sulphur as sulphate (SO_4^{2-}) ions. Sulphur can affect the quality and yield of oilseed crops. Again necrosis of plants may occur due to the S-toxicity.

Functions

1. Sulphur (S) is essential for synthesis of the S-containing amino acids like cystine, cysteine and methionine, which are important for protein synthesis.
2. Sulphur being an element for as a structural formation of chlorophyll in leaves, plays important role in photosynthesis.
3. It is a constituent of proteins and volatile compounds in mustard and vegetables (onion, garlic).
4. The oil synthesis in crops is regulated by S.
5. It is a constituent of Ferrodoxins (Non- Heme iron, Sulphur protein) and S- adenosyl methionine.

Deficiency symptoms

1. Younger leaves become yellowish or light green.
2. Growth is retarded.
3. Fruits become immature.
4. Cell division in plant is affected.
5. Protein quality is impaired for disruption of N-metabolism.

3.6 Micronutrients

These are plant nutrients which are required in a very minute quantities. There are seven micronutrients namely, Iron, Manganese, Boron, Molybdenum, Copper, Zinc and Chlorine.

3.6.1 Iron (Fe)

Iron is present in chloroplasts as a "ferrodoxin" compound. Plants obtain as Fe^{2+} and Fe^{3+} forms and also as chelated -Fe form. Being an immobile element, iron deficiency is noticeable in younger leaves at the growing points.

Functions

1. The biosynthesis of chlorophyll is regulated by Fe.
2. Enzymatic activates are affected by Fe.
3. The oxidation-reduction (Redox) reactions in the plant is being regulated by Fe.

Deficiency symptoms

1. Interveinal chlorosis of younger leaves and on severe deficiency leaves become "Palewhite".
2. Reddish-brown necrotic spots along the leaf margins of young shoots in tree crops.

3.6.2 Manganese (Mn)

It is absorbed by plants as Mn^{2+} form from the soil and is translocated at plant parts as per its requirement.

Functions

1. Involved in redox reactions and electron transport system in photo-system II.
2. It is directly or indirectly involved in chloroplast formation.
3. It acts as a co-factor and catalyses most of the enzymes.

Deficiency symptoms

1. Interveinal chlorosis may be seen on old leaves like that of Iron chlorosis.
2. Depresses inflorescence reducing the leaf and root growth.

3.6.3 Copper (Cu)

A very small quantities of copper are necessary for normal growth of plants. It is absorbed as Cu^{2+}. It is immobile element in plants.

Functions

1. It acts as electron carriers in enzymes influencing the redox reaction in plants.
2. Facilitates utilization of iron in chlorophyll synthesis.
3. The cell wall permeability and nitrate reduction are influenced.
4. The activity of ethylene in fruit-ripening is influenced.
5. Promote the formation of vitamin-A in plants.

Deficiency symptoms

1. Interveinal chlorotic mottling of leaves is observed.
2. The "die-back" (terminal bud wither and die) is most common in fruit-trees.
3. The Cu deficiency induces pollen sterility in microsporogenesis.

3.6.4 Boron (B)

Boron is available at the growing points and in conducting tissue. Plants absorb B as $H_2BO_3^-$, HBO_3^{2-} and BO_3^{3-}. It is immobile element in plants.

Function

1. Essential for cell division in the meristematic tissues.
2. Involved in pollination, pollen formation, pollen tube growth/ flowering and fruit set.
3. Important role in the fertilizing process of plants.
4. It enhances rooting through oxidation process.
5. Acts as a regulator of potassium/calcium ratio in the plant.

Deficiency symptoms

1. Young leaves are deformed, a "rosette" appearance is observed.
2. Premature seed or fruit drop.
3. Browning or hollow stem of cauliflower.
4. 'Heart rot disease' in fruits of the sweet melon (*Cucumis melo*).
5. Tissue break down and prevents sugar and starch accumulation in the leaves.
6. Yellowing of the leaf tip and leaf margin and leaves become scorched.

3.6.5 Molybdenum (Mo)

Required by plants in small quantity and plant absorb as MoO_4^{2-} form. It is the structural component of Nitrogenase enzyme and is a constituent of nitrate reductase.

Functions

1. Takes part in N- Fixation in legumes.
2. Involves in redox reactions in the reduction of NO_3^- to NH_4^+.
3. Bridging action in transferring electrons.

Deficiency symptoms

1. The Mo defiviency may cause reddish or purplish discoloration of leaves, chlorosis and marginal necrosis of leaves.
2. Marginal scorching and rolling or cupping of leaves, "Yellow spot" disease of citrus and "Whiptail" in cauliflower is commonly associated.
3. The NO_3^- accumulation in plants thus inhibits the utilization of N for protein synthesis.
4. Bright yellow mottling between veins, leaves wither.

3.6.6 Chlorine (Cl)

Chlorine can be absorbed easily. The requirement of chlorine is rather quite high as compared to other micronutrients in plants. The role of Cl^- in plant metabolism is still under research. Bronze or yellow colour of leaves with brown or scorched leave margins are found due to the toxicity of Cl^-.

Functions

1. The stomatal regulation (opening and closing) is governed by Cl^-.
2. Water splitting in photo system-II.
3. Bridging ligand for stabilization of the oxidized state of Mn.

Deficiency symptoms

1. Chlorosis and burning at tips and margin of leaves.
2. Yield decreases due to wilting and leaf fall.

3.6.7 Zinc (Zn)

The mobility of Zinc is less in plants. The plants can absorb as Zn^{2+} from soil.

Functions

1. Zn is a constituent of enzymes which regulate various metabolic reactions in the plant.
2. Associated with the formation of some growth hormones in the plant like IAA, and Auxin.
3. Involved in chlorophyll synthesis, protein synthesis in plants.
4. The alcohol dehydrogenase activity in fruit trees is regulated by Zn^{+2}.

Deficiency symptoms

1. Chlorotic and Brown rusty spots on leaves.
2. Rosette appearance in leaf , stunted growth (shortage of internodes).
3. The rate of protein synthesis is reduced and amino acids and amides accumulate.

The sites of deficiency symptoms of plant nutrients (Figure 3.2) vary on the basis of mobility of nutrients in plants.

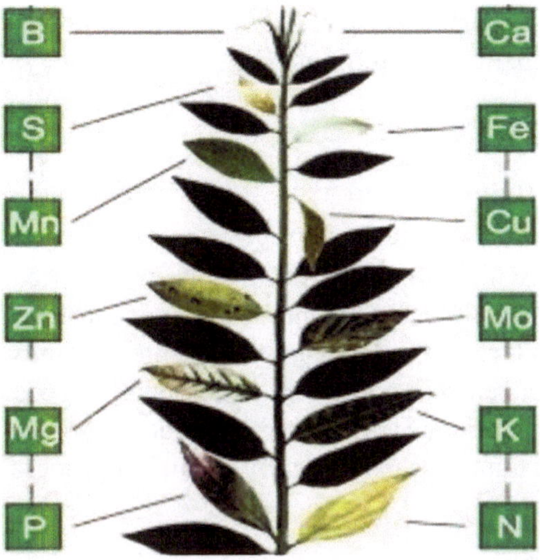

Figure 3.2. Sites of deficiency symptoms of different nutrients in plants

3.7 Transport of Nutrients

Nutrients must get in touch with the plant roots which are to be absorbed by the large number of root hairs. There are three mechanisms by which nutrients can reach the surface of the root hairs. These are: 1) root interception, 2) mass flow, and 3) diffusion.

Nutrient movement from soil to plant: Sometimes, a transportation pull is created within the plant system leading to the development of hydraulic gradient at the vicinity of root from the inner part of the soil. The dissolved nutrients in soil solution can move from the interior of the soil to the root surface, where a concentration gradient may also develop at the rhizosphere. The major processes involving the movement of nutrients from soil to root surface are:

1. **Massflow:** When there is a development of hydraulic gradient between the soil surface and soil, then ions can move from the soil solution as soluble ions (Ca, Mg, Cu, Mn, Mo, NO_3^-, SO_4^{2-} etc.) to the root surface which primarily depend upon the flux rate of water on water consumption of plants. The contribution of mass flow in supplying nutrients to the soil surface depends on soil moisture status and temperature. The flow of water with the soluble nutrients will decrease with the increasing soil-moisture tension. When there is less transpirational demand of plant due to the low atmospheric temperature, massflow becomes less effective. The supply of nutrients by mass flow depends on the supplying capacity of the soil governed by the differences in transpirational rate and nutrient

uptake (varies according to the plant species), soil-water content and availability of nutrients. The quantity of nutrients (FM in kg m^{-2} s^{-1}) is given by

$$\mathbf{FM = qc}$$

where, q is the water flow ($m^3m^{-2}s^{-1}$) and c is the average solute concentration (kg m^{-3}).

2. **Diffusion:** It is the process by which ions or molecules are transported from a higher concentration to a lower concentration (concentration gradient) unless the equilibrium is reached. There is movement of the ions (thermal motion) due to the concentration gradient developed between the bulk soil solution and root surface (lesser nutrient concentration). The nutrients like P and K are mainly transported by diffusion process in soil.

   From the Fick's first law, we get,

   $$dq = -Dp.dc/dx.dt \tag{3.1}$$

   where, dq= amount of diffusing ion across the unit cross-sectional area in time dt.

   dc/dx = Concentration gradient with a –ve sign (as movement of ions is from the higher concentration to lower concentration)

   Dp =is the diffusion coefficient

   dq/dt = Rate of diffusion

   again, $$Dp = Dw.\theta.1/T.1/b \tag{3.2}$$

   Where, Dw = diffusion coefficient in water,

   θ = volumetric soil moisture content,

   T = tortuosity factor

   b = soil buffering capacity

The percentage of clay content, rate of nutrient applied through fertilizer and the placement of fertilizer (band or broadcast) can affect the rate of diffusion in soil. Diffusion is continuous with the length of growing period of plants. If the diffusion rate is low for the high buffering capacity, low soil moisture content or high clay content in soil, then it requires nutrient application in the root zone for subsequent absorption of nutrients by the plants. There will be K^+ accumulation at the inner solution (cytoplasm) of the plant. If the electrical potential difference (E) is -59 mv (at 25°C), the K^+ concentration in the inner solution is ten times higher than that of the outer solution. Thermodynamically, transport against an electro-chemical gradient is called active transport, on the other hand transport down or along an electrochemical gradient is called as passive transport. Addition of metabolic energy is required for the active transport processes.

Hypothesis of ion absorption

Passive absorption: It does not require any addition of energy for transport of ion from the outer solution to inner solution (cytoplasm). Different hypothesis are there, in this regard:

a) The massflow hypothesis: The salts can enter into plants passively along with the water, although the difference in the concentration of salt between the cell and the medium does not qualify the hypothesis.

b) Diffusion: The major driving force here is the difference in the concentration for the net movement of ions. Hence, it requires a constant concentration gradient for the uptake of the ions.

c) Ion exchange: If there is no charge imbalances on both sides of membrane, ions can move across the permeable membrane. There exists a membrane specificity while transporting ion from the outer solution to the inner solution.

d) Donnan equilibrium: At equilibrium, the electrochemical potential of both sides of the membrane would be equal.

i.e., $C_1^+ \times A- = (C_1^+ + C_2^+) \times A^-$ (3.3)

Figure 3.3. Donnan membrane equilibrium system

According to the Donnan equilibrium (Figure 3.3), the concentration of the common ion would be higher on the right hand side than on the left hand side. The system is responsible to deposit either cation or anion in some areas of plant systems but not the both against the concentration gradient.

3. **Root interception:** It is a kind of ion exchange between two solid phases (soil particle and roots) when they are close contact with each other. The roots of the plants may proliferate in the spaces occupied by the nutrients on soil particles. There may be exchange of the H^+ adsorbed at the root surface with K^+ present at the soil surface by contact exchange when their mean oscillation volume overlap each other. The nutrient

absorption by root interception may be increased by mycorrhizal association and effective in relatively less fertile soil.

Nutrient absorption by plants: The absorption of inorganic nutrient is essential for the growth and nutrition of the plants. The mechanism of nutrient uptake by the plants are classified into two categories

i) Active absorption and

ii) Passive absorption

Type of mechanism of ion absorption: The ions can move under the chemical potential gradient or under the electrical potential gradient unless an equilibrium is reached between the outer solution (soil medium) and inner solution (cytoplasm) of the plant and then the electrochemical potential becomes equal on both sides of the membrane. This can be described by Nernst equation. For example, while estimating the electrical charge of inner and outer medium separated by a membrane of an aqueous solution of KCl, then at equilibrium the Nernst equation may be expressed as:

$$\Psi_i - \Psi_0 = E = RT/nF \ln [K^+_0]/[K_i^+] = RT/nF \ln [Cl^-_i]/[Cl_0^-] \quad (3.4)$$

Ψ_i = Electrical charge of the inner medium (cytoplasm)

Ψ_0 = Electrical charge of the outer medium (soil solution)

R = Gas constant, 8.314 JK^{-1} mol^{-1}

T = Absolute temperature in K

F = Faraday constant, 96500 coulombs/equivalent

n =Valence of the ions

Where, E<0, (i.e., cell is –vely charged), the term $[k^+_0]/[k_i^+]$ must be <1, i.e., at equilibrium.

The molecular sieve hypothesis

The cell membrane contain various pore sizes. The ions having smaller ionic diameter can pass through the pores (that is K^+ ion) than the ions having diameter larger than the pore membrane (i.e., hydrated Na^+ ion). Thus, there is higher accumulation of K^+ ions in the cell than the Na^+ ion.

Active absorption: Metabolic energy derived from the aerobic respiration, photosynthesis, glycolysis etc. is required for the uptake of ions by the plants. Several hypothesis are there in this regard.

A. **Lundegardh hypothesis:** Cytochrome, the Fe-containing porphyrin compound present on the outer membrane, act as a carrier of anions and is present both as oxidised (Fe^{3+}) and as reduced (Fe^{2+}) forms. The oxidised cytochrome (cyt^{3+}) at the cell surface can draw an anion (A^-)

from the external medium and form cytochrome-anion-complex (Figure 3.4). After entering into the cell, the cytochrome-anion-complex comes into contact with tonoplast and becomes reduced cyt^{2+} taking an electron from respiratory chain and thus, releases anion. The released anion then combines with another oxidized cytochrome and moves inside the vacuole. Again, the reduced cytochrome releases electron in the outer medium in presence of O_2 and gets oxidized to form cytochrome-anion-complex. The dehydrogenase enzyme splits H_2 into H^+ and e^- and the released e- and H^+ combines with O_2 in the outer medium to produce water. Since, cytoplasm in the cell is negatively charged compare to the outside and hence energy will be spend in pumping anion into cytoplasm and to maintain the electroneutrality, cations will move along with the anion passively. However, this hypothesis had two major drawbacks *viz.*

i) Location of cytoplasm in mitochondria and

ii) Cannot explain the mechanisms of selective absorption of ions.

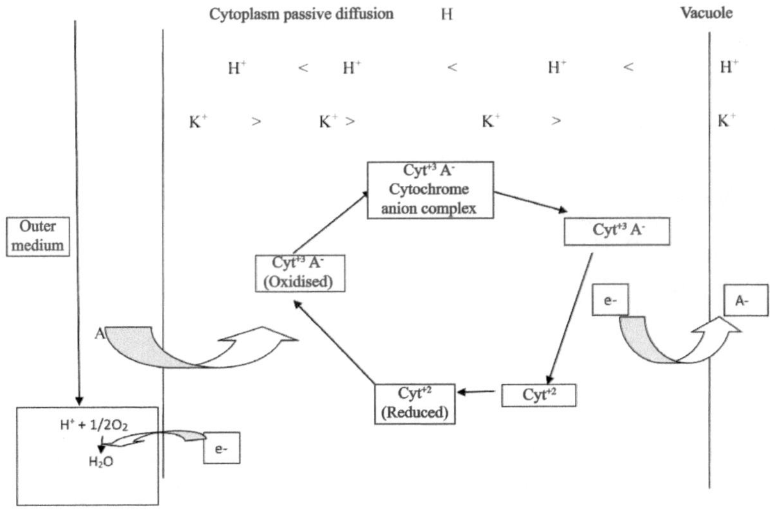

Figure 3.4. Hypothetical ion absorption scheme of Lundegardh

B. Carrier hypothesis: The mechanism of ion absorption is based on the hypothesis that the biological membrane contains molecules, carriers which enable to transport the ions across the membrane. The carriers are specific to bind ions at their active binding sites.

The carrier combines with the ion at the outer membrane boundary to form a carrier-ion-complex that traverses the membrane to phophatase located at the inner membrane boundary. The enzyme phosphatase splits off the phosphate group from the carrier complex (Figure 3.5) on the inner side of membrane and thus, carrier loses the affinity for the ion and thus releases the ion. But, ATP is required for the regeneration of carriers which is mediated by another enzyme called ATP-kinase, located on the inner membrane. The phosphorylated carrier again comes back to the outer membrane boundary and take another ion and the cycle goes on. The ATP may originate from respirartion, photophosphorylation or from glycolysis.

$$\text{Carrier} + \text{ATP} \xrightarrow{\text{Phosphokinase}} \text{Carrier (P)} + \text{ADP}$$

$$\text{Carrier (P)} + \text{ion} \longrightarrow \text{Carrier (P)} - \text{ion complex}$$

$$\text{Carrier (P)} - \text{ion complex} \xrightarrow{\text{Phosphatase}} \text{Carrier} + \text{Pi} + \text{Ion}$$

$$\text{Net : Ion} + \text{ATP} \xrightarrow{\text{Transport}} \text{Ion} + \text{ADP} + \text{Pi} \qquad (3.5)$$

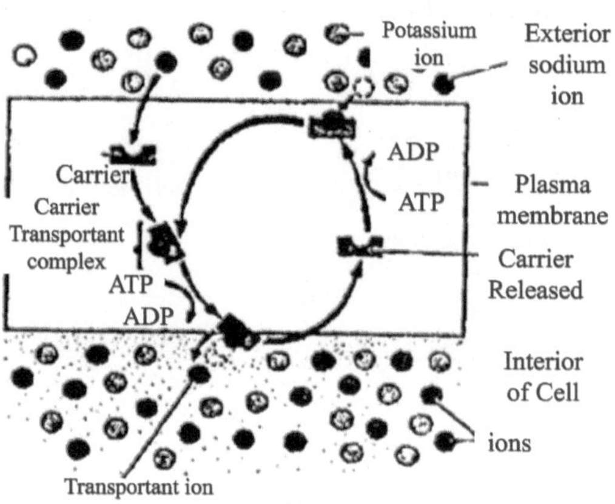

Figure 3.5. Carrier ion transport across membrane

C. **Ion pumps**: It is the reverse process of chemiosmotic ATP synthesis. A membrane-bound ATPase splits ATP into ADP (anion) and a phosphoryl cations $[OP(OH_2)]^+$. On reaction with H_2O and phosphoryl cation, H^+ ion is generated.

$$ATP \rightarrow ADP + [OP(OH_2)]^+$$

$$[OP(OH_2)]^+ + HOH \rightarrow [OP(OH_3)] + H^+$$

The pH gradient is observed (phosphoric acid) across the outer membrane (H^+ pump). A negative electro-potential of the cell is generated with a potential difference between cell and medium in the order of 60 to 160 mv. The ATPase-H^+ pumps are located on plasmalemma and tonoplast and can control pH of cytoplasm (7.3-7.6), apoplasm (□ 5.5) and vacuole (4.5-5.9).

Due to the variation of pH and electrical potential difference, cations are transported passively across the plasmalemma into the cytoplasm along the electrical potential gradient. Movement of solute in a particular direction is called uniport. Cation can move with H+. When the water is moving back to cytoplasm in response of H+ gradient (Figure 3.6) is called co-transport. Co-transport is of two types : i) Symport and ii) Antiport.

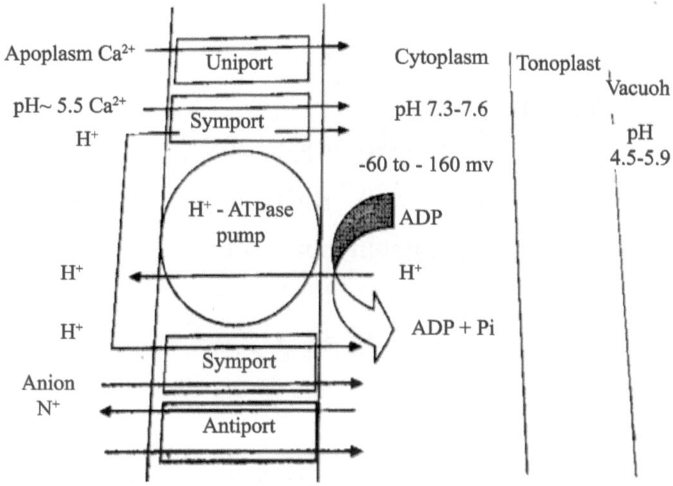

Figure 3.6. Scheme of ion pumps

When the two cations or cation and anion move in one direction is called symport and when, two different ions move across the plasmalemma in opposite direction is called antiport.

Questions

Objective

1. Which of the following ions are common in acidic soils?

   a) Ca^{2+} b) Al^{3+}

   c) Mg^{2+} d) OH^-

2. Which of the following is not an essential nutrient?

   a) Mo b) Cu

   c) B d) Se

3. Which of the following enzymes contain Zn?

   a) Diamine oxidase b) Ascorbate oxidase

   c) Superoxide dismutase d) Plastocyanin

4. Deficiency symptom of which of the following nutrient first appears in older leaves?
   a) Nitrogen b) Copper
   c) Boron d) Iron
5. Which of the following is a beneficial element for plants?
   a) P b) Mo
   c) Na d) Ni
6. Which of the following mineral contains Mn
   a) Sphalerite b) Chalcocite
   c) Pyrolusite d) Borax
7. Most common deficient micro-nutrient in Indian soils is:
   a) Zn b) Mn
   c) Cu d) Fe
8. Which crop deplete maximum micro-nutrients from soils
   a) Rice b) Wheat
   c) Sorghum d) Maize
9. The bacteria involved in nitrification process is
   a) *Nitrosomonas sp.* b) *Desulphovibrio sp.*
   c) *Pseudomonas sp.* d) None
10. Phosphate fixation is prominent in
   a) Acid soil b) Alkaline soil
   c) Saline soil d) Salaine-alkaline soil
11. Khoira disease of rice occur due to deficiency of
   a) Fe b) Mn
   c) Zn d) Mo
12. In soil micronutrient extraction, the pH of 0.005 M DTPA is adjusted to
   a) 7.5 b) 8.3
   c) 8.5 d) 7.3
13. Critical soil test level approach was given by
   a) Cate and Nelson b) Arnon and Stout
   c) Larsen d) Bray

14. The amount of plant available nitrogen in the soil could be decreased by
    a) Ammonification b) Nitrification
    c) Denitrification d) Minerelization
15. Which of the following get oxidised during the first step of nitrification?
    a) Nitrate b) Nitrite
    c) Ammonium d) None of the above

Subjective

1. What are the essential elements for plants in soil ?
2. What are the functional elements ?
3. Differentiate between macro nutrients and micronutrients in soil.
4. What are the tracer elements ?
5. What is the differences between mass flow and diffusion ?
6. Differentiate between the active and passive transport processes in plants.
7. What is critical limit for an ion in a plant ?
8. What is Donnan membrane equilibrium ?
9. What is the deficiency symptoms of nitrogen ?
10. What is the function of P in plants ?

Descriptive

1. What are the sources and forms of plant available nutrients ?
2. What is the criteria of essentiality of an element ?
3. Briefly describe the zone-wise nutrient concentration in plant tissue.
4. Outline the function and deficiency symptoms of potassium
5. Outline the function and deficiency symptoms of sulphur.
6. Outline the function & deficiency symptoms of calcium and magnesium.
7. Describe the functions of cationic micronutrients in plants.
8. Describe the mechanisms of nutrient transport from soil to plant.
9. Outline the process of nutrient absorption by plants.
10. Outline the extractants used for determining the primary, secondary and micronutrient elements from soil.

Suggested Readings

Brady, N.C. and Weil, R.R. (2007). The nature and properties of soils. 14th Edition, Prentice Hall, USA.

Havlin, J.L., Tisdale, S.L., Beaton, J.D. and Nelson, W.L. (2007). Soil fertility and Fertilizers. (Seventh edition), First impression, prentice hall of India Ltd., New Delhi.

Marschner, H. (1995). Mineral nutrition of higher plants, 2nd Ed. (New York; Academic Press).

Mortvedt, J.J., Cox, F.R., Shuman, L.M. and Welch, R.M. (1991). Micronutrients in Agriculture. 2nd Ed. SSSA Book Sewries No. 4. (Madison, Wisconsin: Soil Sci. Soc. Amer.).

4

Soil Organic Matter

4.0 Soil Organic Matter

The soil organic matter (SOM) is a storehouse of multi-nutrient elements, composed mainly of carbon, hydrogen and oxygen and small amounts of other elements like nitrogen, phosphorous, sulphur, potassium, calcium , magnesium and organic residues coming out from the decomposed plant or animal products (Figure 4.1). The SOM may have the recently incorporated materials or age-old decomposed materials or with living flora or fauna in the organic matrix. Soil organic matter is one of the important components determining soil quality indices and also for enriching the fertility status of soil for production maximization. The nutrient retention capacity , water holding capacity, degradation of pollutant elements, carbon sequestration, soil structure and aggregate stability are affected by the presence of SOM.

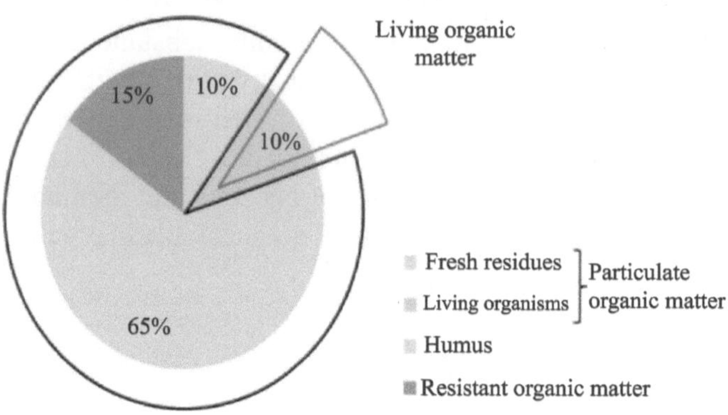

Figure 4.1. Composition of organic matter

4.1 Fractions of Soil Organic Matter

There are four distinct fractions of SOM based on the size, rate of decomposition and composition in soil.

1. Dissolved Organic Matter (DOM) : Mostly it is found in waterbodies as a heterogeneous mixtures which are derived as a product of decomposition of the plant and microbes.

2. Particulate Organic Matter (POM): The POM may be found in the atmosphere due to oxidation of certain organic compounds and when condensed that retain in the filter as course (PM 10), fine particles (PM 2.5) and ultra fine particles (PM 0.1).
3. Humus: Humus is the dark coloured organic substances rich in nutrients, found by the decomposition of plant and animal matter. The peat and muck soil are rich in humus. On fractionation of humus, the humic acid, fulvic acid, hymatomelanic acid and humin are found.
4. Resistant organic substances: The resistant organic substances can withstand elevated temperature, acid and alkaline condition during decomposition. It contains high carbon content (charcoal, coal, graphite). The lignin, derived from the older part of wood is highly resistant for decomposition.

4.1.1 Soil Organic Carbon

Soil organic carbon (SOC) is one of the important components of soil organic matter which varies under the different ecosystems of a region. The carbon sequestration is a challenge to the environment to mitigate atmospheric carbon dioxide and thus, reducing global warming in the sequence. The build-up of carbon in soil largely depends on the agricultural management options and also on the formation of more stable or resistant fractions of SOM.

While understading the process of fractionation of humic substances, the precipitation with acid and subsequent dissolution of part of the precipitated material with alcohol (Figure 4.2) is followed. The fractioned so obtained are the non-descriptive mixture of many chemical compounds. The non-humic substances are isolated by refluxing with a mixture of benzene and methanol (10:1) prior to alkali treatment, since humic substances are insoluble in organic reagents.

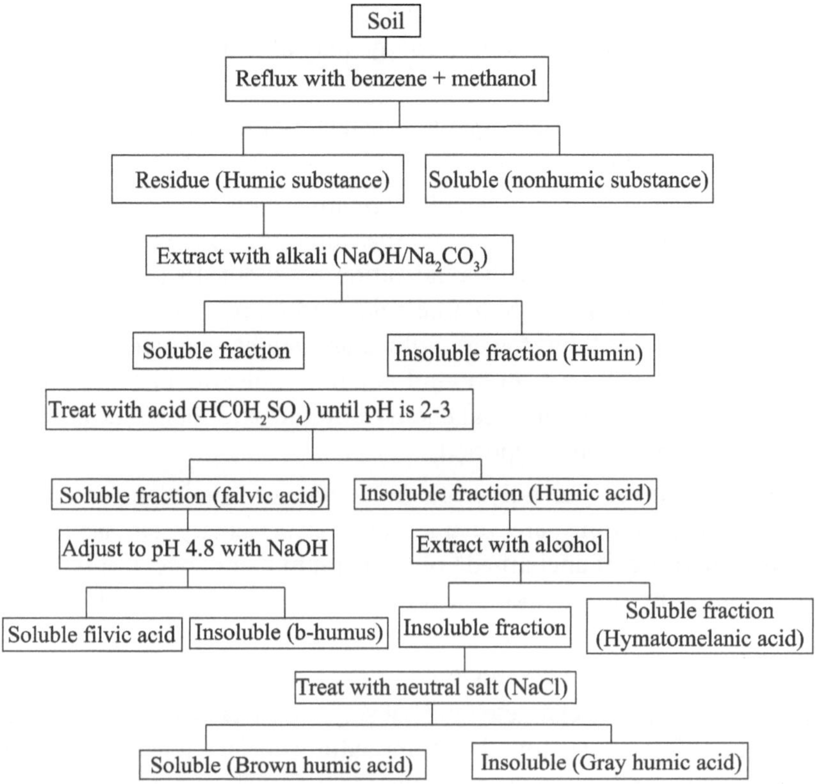

Figure 4.2. Fractionation of Humic substances

4.1.2 Characterization of Soil Organic Fractions

On fractionation of SOM, various organic fractions (Humic acid, Fulvic acid, Humin etc.) are obtained which requires detailed characterisation to understand the processes involved either in carbon sequestration or metal -organo-complexation. The dark-brown, highly coiled polycondensed structure of Humic acid (HA) may have different orientation based on the sources of decomposed plant/animal materials as decomposable substances. In general, the soil humus is a highly complexed colloidal compounds with high molecular weight which can govern the release of most of the ions from the adsorbed to the soil solution phase of a system. The SOM fractionation may broadly be categorized as chemical, physical and biological or biochemical or to some extend morphological. There are different parameters (E_4/E_6, total acidity ,viscosity, molecular weight, surface excess etc.) for characterization of organic fractions to ascertain the turnover rate of organic carbon with the whole soil. The pH-dependent negative charge may also be developed on soil organic matter.

The extracted HA or FA from the organic substances are characterised by the important parameters which are given below:

i) **Shape and size of humic molecule:** The low- molecular weight humic polymer (determined by viscosity measurement) indicated ellipsoid or rod shaped polymers with axial ration of about 10 (Sanyal,2018). The higher molecular weight humic acids had more elongated structure (Visser,1985) than their lower molecular weight counter parts (i.e., Fulvic acid) , which on hydration, the humic molecule become further elongated . The long cylindrical chain of humic acid than 3-D structure or spherical type are also reported.

ii) **Polyprotic nature:** The polyprotic behaviour of humic polymer at its varying concentrations are distinct through viscosity characteristics compared to those of uncharged polymers (at higher concentration). The partial coiling up of HA molecules may be formed due to hydrophobic hydration forming the hydrophobic bonds (Sanyal, 1984).

iii) **Surface tension:** The hydrophobic moieties of humic substances could impart to amphiphilic character and subsequent surface active properties (Tombacz,1999). It was found that, the falling surface tension of HA or FA with concentration could generate positive surface excess (τ) of the humic molecules as found from Gibbs adsorption equation :

$$\tau = -(1/RT).(d\gamma/d\ln C)$$

where, γ is the surface tension of humic solution having the concentration C at absolute temperature T and R the Universal gas constant. The surface excess (τ) has been reported to be increased with the increasing hydrophobicity of the HA molecules (Datta *et al.*, 2001).

iv) **Ratio of elemental composition:** The ratio of C/H, C/N and O/H in humic substances could have impact on the management of cultivation. The practice of cultivation could decrease the C/H and C/N ratio , but in general increase O/H ratio. The C/N ratio decreases due to accelerated polysaccharide mineralization, whereas, the increase in (O/H) indicated the accelerated oxidation and humification (Martin *et al.*,1998). The decrease of (C/H) could be due to the accelerated mineralization of lignin under cultivation.

v) **Spectroscopic property :** The spectroscopic properties of HA and FA is characterised by measuring the ratio of optical density (O.D.) of dilute aqueous HA or FA solutions at 445 and 665 nm (E_4/E_6) to ascertain the degree of condensation of aromatic and aliphatic groups (Schnitzer and Khan,1972). Besides, the Infrared (IR) spectroscopic techniques are used for functional group analysis and also to detect structural alterations in soil - humic substances.

4.1.3 Organo – metal complexes

To understand the metal -organic or organo-metallic complexation reaction in a system, the respective properties of the metal and also of the organic moiety is required, where there should exist chemical compounds containing at least one chemical bond between a carbon atom of an organic molecule and a metal. There may also have the organyl fragments or molecules, bonds to 'inorganic' carbon, like carbon monoxide to form the complex .In other words, "metal-organic compound" may be assigned to metal-containing compounds which may not have direct metal-carbon bonds but may have linked with organic ligands.

The concept of organic metal complexation is applied to form Organometallic compounds (Figure 4.3) which are being used in research and industrial sectors. However, when any ligand form a direct M-C (Metal-Carbon) bond, then complex is usually considered to be organometallic, e.g., $[(C_6H_6)Ru(H_2O)_3]^{2+}$. Formation of transition metal alkyl complex is methylcobalamin (a form of Vitamin B_{12}), with a cobalt-methyl bond. The Zeise's salt is an example of a transition metal alkene complex.

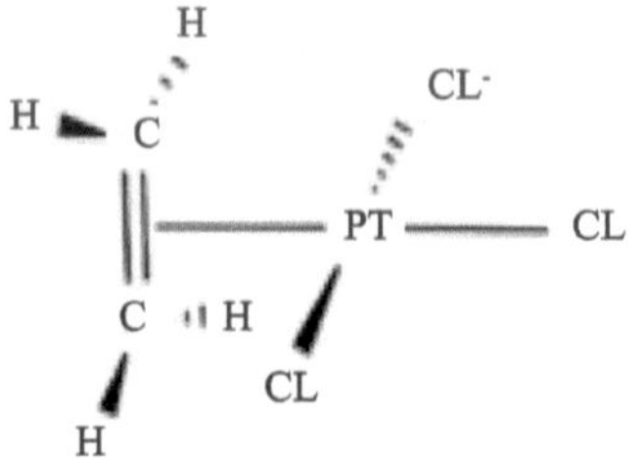

Figure 4.3. Formation of Organo-metal complex

The organo-metal interaction to a large extend depend upon the nature of the mineral ,pH and in some cases on the concentration of the background electrolyte (Yuan *et al*,2000). The mineralogy of the clay (expanding/non-expanding) along with agro- climatic situation often influence the activity of the oxides in SOM stabilization. The functional groups of humic acid (enolate, heterocyclic N, phenolic -OH, carboxyl group, sulfhydryl) also take the role in binding ions for stability of the complex. Sometimes, the basal surface and edges of the clay (Illite) and or metal ions are associated for bonding with humus. It was observed that, the fixation of FA by clay minerals (at pH 2.0) was in the order Illite > Kaolinite>Montmorillonite . The more negative charge density on FA over that of HA and hydrophobicity might account for weaker complexation of FA by the given clay minerals.

Besides, the stability constant (logK) of metal -humic complexes are also important to understand the nature of humic materials while retaining the metal ions in their matrix. The relative proportion of aromatic and aliphatic carbon in humus are also important in releasing ions from the complex. Besides, the soil clay -organic complexes are also associated with exothermic peaks around 340^0,435^0 and 510^0 C (Manoj Kumar, 1995) as ensured by the oxidation of organic matter, complexed to clay minerals surface, sesquioxide surface and allophane respectively.

4.1.4 Decomposition of Organic Matter

The organic matter undergoes aerobic and anaerobic decomposition where complex organic substances are broken down into simpler organic fractions with the progress of time. It is a natural process where the living organisms are decomposed shortly after their death. The decomposing organisms (microorganisms) are responsible to facilitate biological action in the process of decomposition. The techniques of recycling the organic waste back in its usable shape is now being practiced to safeguard the environment. In doing so, the organic materials (vegetable , animal manure and other organic substances) are being converted from wasted materials to a more stable form for further use as a soil amendment. This process is known as composting and the final product of composting is called compost .

Anaerobic

Sugar $\Longrightarrow$ Alcohol + CO_2 + Energy

Aerobic

Sugar + O_2 $\Longrightarrow$ CO_2 + Energy

The major driving forces responsible for the decomposition of SOM are bacteria, fungi, molds, protozoa, actinomycetes, and other saprophytic organisms at the initial stages. In the later stages, the decomposition is governed by the mites, millipedes, centipedes, springtails, beetles and earthworms for further fragmentation for enriching the composting materials. Depending upon the available microclimate in the pile, there is variation on the type organisms. The presence and absence of oxygen in the microclimate results two types of decomposition processes (Anaerobic and Aerobic decomposition).

4.1.4.1 Anaerobic Decomposition

Anaerobic decomposition occurs in absence of oxygen and there is breakdown of organic material anaerobically through putrefaction. The organic compounds are broken down by the action of anaerobic organisms in the normal case. These organisms utilise nitrogen, phosphorus and other nutrients to develop cell protoplasm and also reduce the organic - nitrogen to organic acids and ammonia. The carbon from the organic compounds which is not utilized in the cell protein is liberated mainly in the reduced form of methane (CH_4) and partly as carbon-dioxide through respiration. The product developed through the anaerobic decomposition of organic substances when subject to exposure in air for application purpose, further decomposition may start aerobically. It is observed that less heat is generated in anaerobic decomposition. In anaerobic process there is slower rate of disappearance of pathogenic organisms in the organic mass and more likely to comeback as fatal parasites in the waste if there is incomplete decomposition.

4.1.4.2 Aerobic Decomposition

When organic substances are decomposed in presence of oxygen, the process is called aerobic decomposition . The aerobic process is most common in the natural ecosystem to form a relatively stable compounds with no obnoxious odour in it. The living organisms can utilise the nitrogen, phosphorus, carbon and other nutrients as its food materials in presence of O_2. The available carbon may serve as a source of energy for the organisms and is respired as carbon dioxide (CO_2) for which much more carbon than nitrogen is required for the development of cell protoplasm. There may be requirement of much of microbial populations if the excess of carbon over nitrogen (C:N ratio) in organic materials are there and also to burn most of the carbon as CO_2. The aerobic organisms again thrive well on the dead tissues or cells of the organisms by using nitrogen from the dead tissues and liberating CO_2 and thus balancing the C:N in the system. During the decomposition , much of the heat is generated (~ 70kcal/gmole of glucose) under the aerobic condition and in some cases a rise in temperature to the tune of 170^0F was observed during the aerobic decomposition of organic substances. Some of the mesophilic and thermophilic bacteria are involved in the process of aerobic decomposition when there is exothermic reactions leading to the temperature variation from 0^0F to 160^0F. The rate of decomposition is rapid at the thermophilic temperature under oxidation over that of mesophilic temperature. The odourless product after the decomposition through the intermittent supply of oxygen into the pit is the indication of the complete aerobic oxidation of organic substances. A fairly stable humus can be obtained through aerobic decomposition.

4.1.5 Factors Affecting Rate of Decomposition of SOM

The rate of decomposition depends by and large on microbial activity in association with aeration and sufficient food materials for the organisms (Figure 4.4).

Some of the factors affecting the rate of SOM decomposition are :

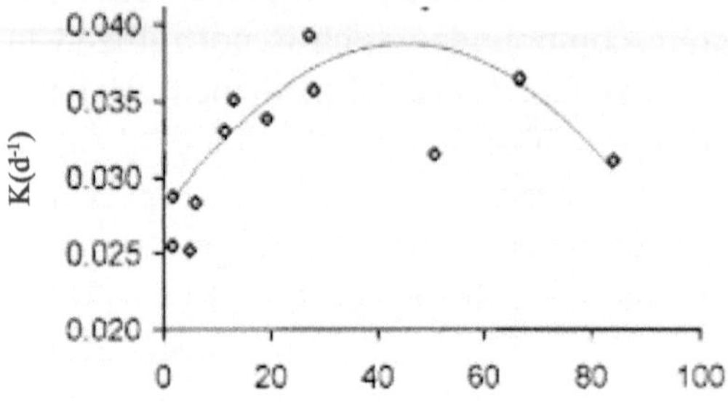

Figure 4.4. Relationship between decomposition rate (K) and percentage of organic matter in sediments.

i) **Aeration** It requires oxygen for the microbial decomposition of organic matter. When the decomposition occurs in absence of oxygen (anaerobic conditions) the process becomes slow and foul odours may develop. Hence, there should be flow of O_2 (10-12%) periodically in a decomposing pit in a residential setting to hasten the composting process.

ii) **Moisture** To continue the microbial decomposition in a compost pit/pile ,proper moisture is to be maintained periodically in such a way that, the pile is damp but not soggy. If the pile is too wet, it should be dried out first and then moistening can be done.

iii) **Particle Size** The decomposition time is reduced on grinding the organic materials before composting. The smaller the size of the organic particles, the more quickly it can be assimilated by the microbes.

iv) **Temperature** The rate of decomposition is enhanced by the increase in temperature. The microbes that make up the bulk of the decomposition process fall into two categories: mesophilic (those that live and function in temperatures of 50 to 113°F), and thermophilic (those that thrive at temperatures between 113 to 158°F). A well-mixed, adequately working compost pile will heat to temperatures between 110°F to 160°F as the microbes actively feed on the organic materials.

v) **Organisms** There are microscopic as well as visible organisms which take part actively in the decomposition process. Compostable organic materials normally contain a large number of bacteria, fungi, molds, and other living organisms.

vi) **Chemical Decomposers** The chemical decomposers are the initial inhabitants of the pile which make the environment of the pit a favourable condition so as to start their normal functions of breaking down organic materials into functions.

vii) **Bacteria** Bacteria likely to be found in a compost heap are aerobic bacteria that break down organic compounds and thrive in temperatures ranging up to 170°F (77°C). Bacterial populations (aerobic or anaerobic) differ from pile to pile, depending upon the raw materials of the compost, degree of heat, amount of air present, moisture level, geographical location of the pile, and other considerations. A decrease in the temperature of the pile or a sharp change in its acidity can render bacteria inactive or kill them. When the environment of a heap begins to change, bacteria that formerly dominated may be replaced by another species.

viii) **Actinomycetes** When there is declining of the bacterial population, there is growth of actinomycetes in the compost pile which have the ability to produce antibiotics, chemical substances that inhibit bacterial growth.

ix) **Protozoa** Protozoa are the single - celled microscopic organisms and are more complex in their activities than most of the bacteria. Protozoa obtain their food from organic matter in the same way like bacteria, but play a much smaller part in the composting process, because of the less population in respect to bacteria.

x) **Fungi** Fungi are multi-celled, filamentous or single-celled primitive plants which can take part in the decompositions process in the compost heap around 70° to 75°F, though some thermophilic forms prefer much greater heat and survive at 120°F. The mites, millipede , centipedes , snails and slugs are directly and indirectly involved in the decomposition process.

xi) **Earthworms** There are so many beneficial effects of the activities of earthworm to sustain the fertility status of soils. The elementary canal of the earthworm can ingests, decomposes, and deposits casts continually during the earthworm's active periods. As soil or organic matter is passed through an earthworm's digestive system, it is broken up and neutralized by secretions of calcium carbonate from calciferous glands near the worm's gizzard where the material is finely grained prior to digestion. Digestive intestinal juices rich in hormones, enzymes, and other fermenting substances continue the breakdown process. The matter passes out of the worm's body in the form of casts, are the richest and finest quality of all humus material. Fresh casts are markedly higher in bacteria, organic material, available nitrogen, calcium and magnesium and available phosphorus and potassium than soil itself. Earthworms thrive on compost and contribute to its quality through both physical and chemical processes, and reproduce readily in the well-managed pile. Hence, earthworm which can participate in compost- making should be nurtured carefully for improving the fertility status of soil and ultimately the production potential of the country and thus, earthworms are called the Farmers' friend in nature.

4.2 Carbon Cycles

The carbon cycle is the process of recycling carbon atoms in nature. The Carbon is considered as the foundation for all life on Earth. The scientists consider 99.9% of all organisms on the planet to be carbon - based life for survivality and is needed as carbon in the form of a sugar or carbon dioxide gas in the environment. The carbon is continuously cycled and reused. Considering the present stock of carbon on earth , it (carbon) requires to be recycled in either form through the rapid carbon exchange among living organisms and in the way of long-term cycling of carbon under geologic processes. The carbon cycle may be subdivided into biological carbon cycle and geological carbon cycle involving organisms and physicochemical actions respectively.

4.2.1 Biological Carbon Cycle

The **autotrophs**, or self-feeders (plants or algae) capture carbon dioxide from the air or bicarbonate ions from the water and use them to make organic compounds such as glucose and thus there is entry of carbon in the food webs. The h**eterotrophs,** or other-feeders, such as humans, consume the organic molecules, and the organic carbon is passed through food webs. The energy stored in carbon-containing molecules are released by the actions of autotrophs and heterotrophs on breaking down the molecules in a process called cellular respiration and the carbons of the molecule are released as carbon dioxide. The organic compounds and carbon dioxide are also released on break down of dead organisms and waste products by the decomposers. Overall, an estimated 1,000 to 100,000 million metric tons of carbon can move through the carbon cycle by biological pathway each year.

4.2.2 Geological Carbon Cycle

The carbon cycle through the process of the geological pathway takes longer time than the biological pathway to the tune of millions of years for carbon to cycle through the geological pathway. Carbon may be stored for long periods of time in the atmosphere, water bodies, ocean sediment, soil, rocks, fossil fuels, and Earth's interior.

The level of carbon dioxide in the atmosphere is influenced by the reservoir of carbon in the oceans and vice-versa. The atmospheric carbon dioxide dissolves in water and reacts with water molecules in the following reactions:

$$CO_2 + H_2O = H_2CO_3 \rightarrow H^+ + HCO_3^-$$

It is found that, carbon is stored in soil as organic carbon from the decomposition of living organisms or as inorganic carbon from weathering of terrestrial rock and minerals. The fossil fuels (oil, coal, and natural gas), the remains of plants are decomposed under anaerobic conditions are found underground and carbon is released into the atmosphere as carbon dioxide from these fossil fuels on burning. Besides, the carbon can enter the atmosphere by the eruption of volcanoes. Carbon-containing sediments in the ocean floor are taken deep within the Earth in a process called subduction, in which one tectonic plate moves with another and carbon dioxide formed by this process can be released into the atmosphere by volcanic eruptions or by the hydrothermal vents.

Thus, in a carbon cycle there are valid sources of carbon (soil, plant and animal residues) on earth and in atmosphere which on decomposition by the autotrophic and heterotrophic organisms release carbon -di -oxide in the atmosphere (Figure 4.5). Some of the carbon is immobilised by the microorganisms and some are being stored in as fossil fuels for subsequent utilization.

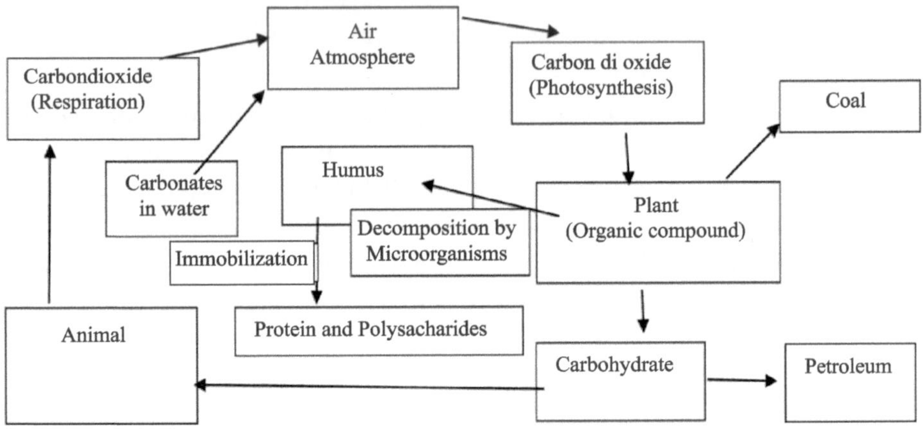

Figure 4.5. Carbon cycle

There is a gradual demand on earth's limited fossil fuel reserves since the beginning of the Industrial Revolution. Fossil fuels are considered a non-renewable resource which are being used up at a faster rate than they are produced by geological processes. Hence, the natural processes, such as volcanoes, plant growth, soil carbon levels, and respiration should be taken into account for predicting the impact of the carbon stock on utilising fossil fuels and on global climate change.

4.2.3 Soil carbon fractions

On fractionation of soil carbon pools , one can predict the availability and lability of carbon in the environment. The important C fractions are given below:

i) **Total organic carbon (TOC)**: It is the amount of carbon found in an organic compound and can act a non-specific indicator of water quality or cleanliness of pharmaceutical manufacturing equipment. The TOC sometimes refer to the organic carbon in soil. The total carbon (TC) measures both the total organic carbon present as well as the inorganic carbon (IC), the latter representing the content of dissolved carbon dioxide and carbonic acid salts. By subtracting the inorganic carbon from the total carbon yields TOC. Another common variant of TOC analysis involves removing the IC portion first and then measuring the leftover carbon. This method involves purging an acidified sample with carbon-free air or nitrogen prior to measurement, and so is more accurately called non-purgeable organic carbon (NPOC).

ii) **Microbial biomass carbon** (MBC) **:** Microbial biomass carbon is a measure of the carbon (C) which is found within the living component of soil organic matter (i.e., bacteria , fungi etc.) while decomposing the soil organic matter by the microbes, where, carbon dioxide and plant available nutrients are released.

iii) **Particulate organic carbon (POC) :** Particulate organic matter (POM) is the organic fractions in 53±2000 ɥm soil separates (Cambardella & Elliott 1992; Wilson *et al.* 2001) of which the carbon content is referred to as particulate organic carbon (POC).

iv) **Labile carbon :** Labile carbon is the fraction of soil organic carbon with most rapid turnover times and its oxidation drives the flux of CO_2 between soils and atmosphere. Labile carbon mainly consists of soil microbial biomass carbon, dissolved organic matter, and easily oxidative organic matter.

v) **Recalcitrant carbon**: The recalcitrant carbon usually refers to the component of SOM that is resistant to microbial decomposition or protected by mineral of soil particles.

4.2.4 Management of soil carbon

The SOM is in a constant state of turnover, where it is decomposed and replaced by new organic material. Therefore, the balance between inputs and the rate of loss will determine the relative flux in SOM content. The declining trend of SOM is still a major concern and thus current land management practice is typically taking more from the system than it returns. Followings are some of the important measures that may be taken into consideration to improve the soil organic carbon content in soil.

i) To introduce zero or minimum tillage for addressing minimum disturbances in soils.

ii) To include crop rotation in a sequence.

iii) Adopting proper management strategies for improving soil structure and maximising organic matter returns.

iv) To minimise wind erosion , stubble should cover a minimum of 70% of the soil surface, preferably be standing.

v) To grow high yielding crops with high biomass and pastures, and to maximise crop frequency to increase organic matter returns to the soil.

vi) A combination of inorganic and organic fertilisers to maximise crop production.

vii) Manure/compost or other organic amendments may be incorporated .

viii) Periodical monitoring to assess the replenishment and reserve of the carbon .Due attention should be given to increase soil carbon to enhance productivity first, rather than to focus on the potential of carbon as a financial stream in carbon markets.

ix) Perennial plants may be grown to provide a constant C input in soil.

x) Proper recycling of residues (plants/animals)in soil.

xi) The return of manure and recycled organic materials to the soil will be considered for potential increase of SOC levels.

xii) Carbon sequestration into the stable SOC pool requires predictable amounts of N, P and S and that carbon sequestration will be limited when these nutrients are insufficient despite large amounts of carbon input.

xiii) The stubble retention without changing elsewhere in management practices will contribute to the maintenance of soil carbon levels, rather than an increase. This will also contributes to a more diverse microbial community.

Questions

Objective

1. The main component of biogas is
   a) CH_4 b) CO_2
   c) SO_2 d) N_2O
2. Rate of organic mater decomposition is increased by
   a) Temperature b) Moisture
   c) Texture d) None of the above
3. For mechanical analysis of soil, organic matter is destroyed by heating the soil with
   a) H_2O_2 b) NaOH
   c) $K_2Cr_2O_7$ d) HCl
4. Organism that need oxygen for respiration and cannot survive without it is called
   a) Obligate anaerobes b) Obligate aerobes
   c) Facultative anaerobes d) None of the above
5. Which epipedon is not common in Indian soils?
   a) Mollic b) Ochric
   c) Umbric d) Folistic
6. Which one of the following is not a soil organic horizon
   a) Mor b) Mull
   c) Plinthite d) Orterde
7. Soil structure is degraded due to
   a) Loss of organic matter b) Adition of fertilizer
   c) Cultivation of leguminous crops d) Addition of organic matter
8. Carbon dioxide content in soil is
   a) 0.3% b) 0.03%
   c) 3.0% d) None of the above
9. The formative element 'od' stands for
   a) Mollisol b) Aridisol
   c) Oxisol d) Spodosol

10. The colour ratio of E_4/E_6 value of fulvic acid is
    a) 1-3 b) 3-5
    c) 5-7 d) 7-8
11. Humic acids are amphoteric and .... in nature.
    a) Monobasic b) Polybasic
    c) Monoacidic d) Polyacidic
12. Method that analyses the unpaired electron spins in paramagnetic organic materials is
    a) Infrared spectroscopy
    b) Electron paramagnetic resonance
    c) Nuclear magnetic resonance
    d) Magnetic resonance spectroscopy
13. Materials used in gel chromatography as the gel substance in the study of humic materials are
    a) Polysaccharides and polystyrenes
    b) Polyamides and Al-oxides
    c) Cellulose, agar and glass beads
    d) All
14. Generally, polysaccharides of fulvic acid are less branched than those of humic acids (True/False)
15. What is the conversion factor for the conversion of nitrogen into protein?
    a) 6.25 b) 6.724
    c) 6.75 d) 6.50

Subjective

1. Outline the concept of soil organic matter (SOM).
2. What are the functions of SOM ?
3. What are the parameters for characterising SOM ?
4. What are aerobic and anaerobic deçomposition ?
5. What is the effect of temperature on soil organic matter decomposition ?
6. What are the fractions of soil carbon ?
7. What are the decomposers of soil organic matter ?

8. What are the labile and non-labile carbon in soil ?
9. How does SOM affect the aggregate stability of soil ?
10. Why SOM is called the storehouse of nutrients ?

Descriptive

1. Outline the importance of SOM in agriculture.
2. Describe the formation of organo-metal complexation in soil.
3. What are the different processes of organic matter decomposition ?
4. What are the factors affecting the rate of SOM decomposition ?
5. Describe the parameters for characterising the SOM fractions.
6. What are the organisms responsible for SOM decomposition ?
7. Describe the carbon cycle.
8. Describe in brief the soil carbon pools.
9. How the SOC stock is improved ?
10. How does the global climatic change affect the SOC in the environment?

Suggested Readings

A text book of Soil Chemistry. S.K.Sanyal, (2018). Daya Publishing House, Division of Astral International Pvt Ltd. New Delhi 110002.

Datta, A., Sanyal, S.K. and Saha, S. (2001). A study on natural and synthetic humic acids and their complexing ability towards cadmium. Plant and Soil, 235: 115-125.

Manoj Kumar (1995). Ph.D. thesis IARI, New Delhi, cited in Ghosh and Ghosh (1998).

Martin, D., Srivastava, P.C., Ghosh, D. and Zech, W. (1998). Characteristics of humic substances in cultivated and natural forest soils of Sikkim. Geoderma. 84: 345-362.

Sanyal, S.K. (1984). Structure of water in solution of organics-Hydrophobic hydration. Chem. Edn. (UGC). 1(2): 14-18.

Schnitzer, M. and Khan, S.U. (1972). Humic substance in the environment, Marcel Dekker, New York.

Tan, K.H. (2005). Principles of Soil Chemistry. Marcel Dekker, Inc. New York.

Tombecz, E. (1999). Colloidal properties of humic acids and spontaneous changes of their colloidal state under variable solution conditions. Soil Science. 164: 814-824.

Varadachari, C., Mondal, A.H., Nayak, D.C. and Ghosh, K. (1994). Clay humas complextaion ; Effect of pH and the nature of bonding. Soil Biol. Biochemistry, 26: 1145-1149.

Visser, S.A. (1985). Viscometric studies on molecular weight fractions of fulvic acids and humic acids of aquatic terrestrial and microbial origin. Plant Soil, 87: 209-221.

Yuan, G., Theng, B.K.G., Parfitt, R.L. and Percival, H.J. (2000). Interactions of allophane with humic acid and cations. European Journal of Soil Science, 51: 35-41.

5

Soil Colloids

Soil colloidal substances are composed of very fine particles (< 0.002 mm diameter) that undergoes two phase system in a heterogenous manner, where, one phase (disperse phase) is disperse in a fine state of subdivision in another medium called dispersion medium. The soil colloidal properties are affected by the physico-chemical and biological parameters of soils. The availability of plant nutrients are governed mostly by the surface charge characteristics of clay colloids.

5.0 Soil Colloids

Soil colloids being one of the active constituents determine the physical and chemical properties of the soil with the diameter ranging from 1 μm to 2 μm. The origin of the particles may be organic and inorganic constituents.

5.1 Importance of Soil Colloids

The surface area of contact of the colloidal particles with ions (particularly with the positively charged ions) could determine the transport of nutrients needed for plant growth and nutrition. When the ions like Calcium (Ca^{++}), Magnesium (Mg^{++}), Potassium (K^{+}), and Sodium (Na^{+}) etc. are dissolved in a soil-water solution, become available to plants when they are in close contact with root surfaces. The adsorption capacity of soil colloids restricts the important nutrients to be leached out from the soil and thus becomes available to the plants when in need at the favourable reaction atmosphere. The soil colloidal particles holding the maximum amount of ions are considered to be more fertile. The humus which are also considered as a colloidal substance can hold ions required for the plant growth. However, replacement of the adsorbed ions from the soil colloidal particles depends on the physico-chemical properties of the matrix and the presence of any complementary and competing ions for the same adsorption sites.

5.1.1 Properties of Soil Colloids

Following are some of the important properties of soil colloids:

i) Colloidal particles are negatively charged particles and are in dynamic motion.

ii) Addition of ions of oppositely charged may transform the colloidal particles into separate state (i.e. liquid to semisolid state etc.).

iii) Colloidal particles have ability to absorb gases, liquid and solid from their suspension.

iv) Colloidal particles are very specific while passing through a semi-permeable membrane.

v) Brownian movement: Colloidal particles are found to be in random and zig-zag movements, colliding each other and again comeback to its original position.

vi) Flocculation: The colloidal particles are coagulated by adding an oppositely charged ion. The flocculation refers to the process by which fine particulates are clubbed together into a floc. The floc may then float to the top of the liquid, settle to the bottom of the liquid (sedimentation), or may be filtered from the liquid.

vii) Electrical charge: Charge on Colloidal Particles. The colloidal particles carry an electric charge. The colloidal solution having the suspended particles possess either positive or a negative charge.

viii) Adsorption: The adsorption is a surface phenomena. Typical colloidal particles attach to fluid–fluid interfaces, composed of relatively polar and non-polar fluid phases. The particle adsorption to the interface decreases the interfacial tension due to the reduction in the contacting interfacial areas between the two fluids. The adsorption process is guided by two empirical formulas.

1. Langmuir adsorption isotherm

   $x/m = ABC/(1+BC)$

2. Freundlich adsorption isotherm

   $x/m = kC^{1/n}$

where, x/m is the amount of ions adsorbed per unit amount of clay

A and B are constants. C= equilibrium concentration of the solution.

k = Freundlich constant. n = constant (n>0)

ix) Cohesion and adhesion: Unlike sand, clay particles possess the properties of cohesion. The colloidal particles can exhibit the cohesion and adhesion behaviour in the matrix.

x) Swelling and shrinkage: Shrinking and swelling are generally associated with fine-grained clay soils. This increased surface tension tends to pull adjacent soil particles come closer. Colloidal particles (percent less than 1 μm) have the property of swelling and shrinkage.

xi) Plasticity: It is the capacity of being moulded. The Colloidal particles can show the properties of plasticity depending upon the amount of clay content.

5.1.1.1 Concept on Coagulation, Precipitation, Flocculation & Peptization

- Coagulation: The destabilization of colloids by neutralizing the electric charge of the dispersed phase particles resulting in aggregation of the colloidal particles is called Coagulation. When the charges are neutralized the particles may collide and bind to each other.
- Precipitation: The phenomenon of precipitation of a colloidal solution by the addition of the excess of an electrolyte is called coagulation or flocculation." This reduces the charge on the particles and ultimately they settle down to form a precipitate.
- Flocculation: It is a process whereby small particles in suspension are caused to aggregate, giving large clusters (flocs) that are more easily separated than the original particles. The process is in use in many industrial applications.
- Peptization: It is the process for the formation of stable dispersion of colloidal particles in the dispersion medium. It may also be defined as a process of converting a precipitate into a colloidal sol by shaking it with dispersion medium in presence of a small amount of electrolyte. When the ferric chloride is added to the precipitate of ferric hydroxide, the hydroxide precipitate transfer to the sol by absorbing ferric ions.

5.1.2 Types of Soil Colloids

Depending upon the structure, composition and properties, soil colloids can be classified as follows:

Soil colloids

A. Inorganic colloids
   a) Crystalline Silicate clays (Illite, Kaolinite)
   b) Non-crystalline (Allophane)
   c) Iron and aluminium oxides of clay (Gibbsite)
   d) Mixed or interstratified layer clay
B. Organic colloids, e.g., (Humus)

5.1.2.1 Crystalline silicate clays

(Phylosilicates → tetrahedral and octahedral crystal sheets)

1. **Tetrahedral Sheet:** Tetrahedral sheets are composed of individual tetrahedrons which share every three out of four oxygens. They are arranged in a hexagonal pattern with the basal oxygens linked and the apical oxygens pointing up/down.

   Silicate mineral structures is the silicon-oxygen tetrahedron $(SiO_4)^{4-}$. It consists of a central silicon atom surrounded by four oxygen atoms in the shape of a tetrahedron (Figure 5.1).

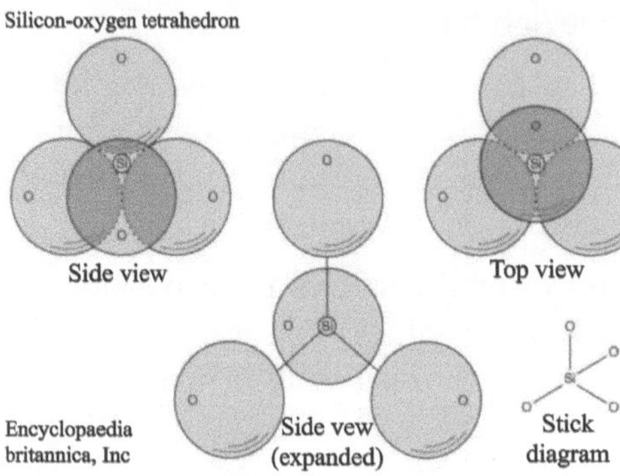

Figure 5.1. Silicon tetrahedral model

2. **Octahedral sheet**: Octahedral sheets are composed of individual octahedrons that share edges composed of oxygen and hydroxyl anion attached with Al, Mg, Fe^{3+} and Fe^{2+}. These octahedrons too, are arranged in a hexagonal pattern.

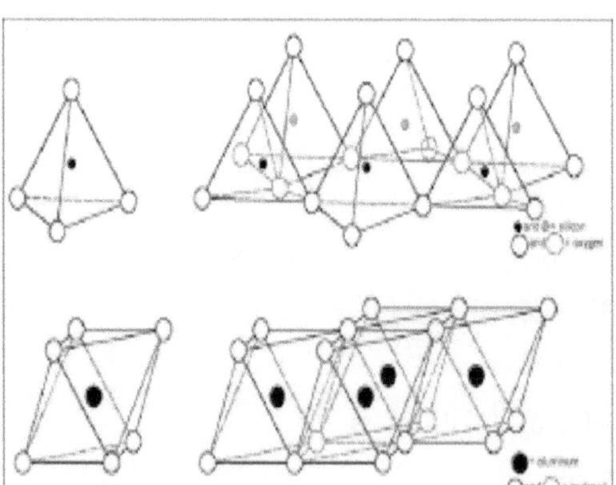

Figure 5.2. Structure of silica tetrahedral sheet

Goldschimidt's Laws of Crystal Structure

The ratio of the number of cations and anions can govern the crystal structure to a large extend. Even the repulsive forces exerted by the ionic compound specifically when the cation size is small and anions are in virtual contact with one another are observed . While understanding the structure of phyllosilicates, the laws of Goldschimidt are employed.

Law-I: In an ionic crystalline compound , isomorphous replacement of one cation by another , without incurring any change in the order of the crystal pattern is permitted provided that the radii of the cation replaced and the cation replacing agrees within 10-15%.

The isomorphous substitution may occur in sheets and charge imbalances may be created in the crystal if the valance of two cations concerned are different.

Law-II: When the anions surround a central cation (in a closed packed manner) to satisfy its charge completely , the number of anion that can be so accommodated around the central cation depends on the ratio of the radius of the cation to that of the surrounding anion. This ratio basically govern the number of surrounding anions about a central cation or the coordination number of the central cation.

The coordination number may be defined as the number of nearest neighbouring ions that surround one ion of opposite charge in order to attain maximum stable (minimum energy) compact neutral crystal structure.

Structure of crystalline silicate clays

The most important silicate clays are alumino-silicates or phyllosilicates originated from the primary silicate minerals.

1. **Kaolinite group (1:1 type silicate)**;There are four species in this group– Kaolinite, Halloysite, Dickite and Nacrite. The adjacent OH- plane of octahedral sheet in one layer is bound by the H-bonding (Figure 5.3) with the oxygen plane of tetrahedral sheet of the next layer. It is because of the inter layer H-bonding , the structure of Kaolinite is fixed and there is no expansion of this mineral in presence of polar liquid or water. The basal spacing along c-axis of kaolinite is fixed at 7.2 A^0 and the specific surface area is 7-30 m^2/g.

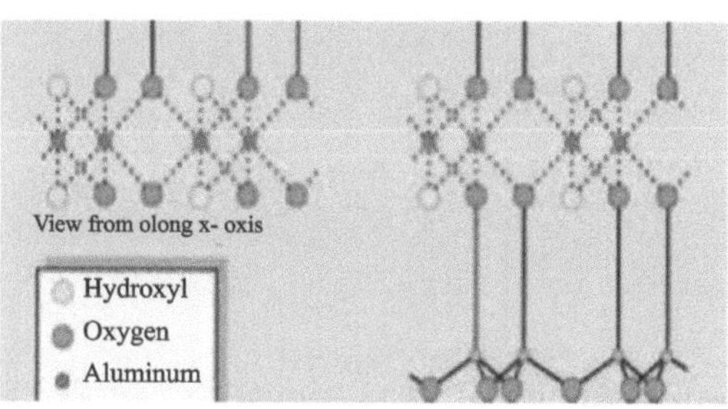

Figure 5.3. Structure of Kaolinite

2. Lattice type of 2:1 silicate clay; In 2:1 type of mineral, the octahedral layer is fixed in between two adjacent tetrahedral layers (Figure 5.4). This groups are having expanding and non-expanding type of minerals.

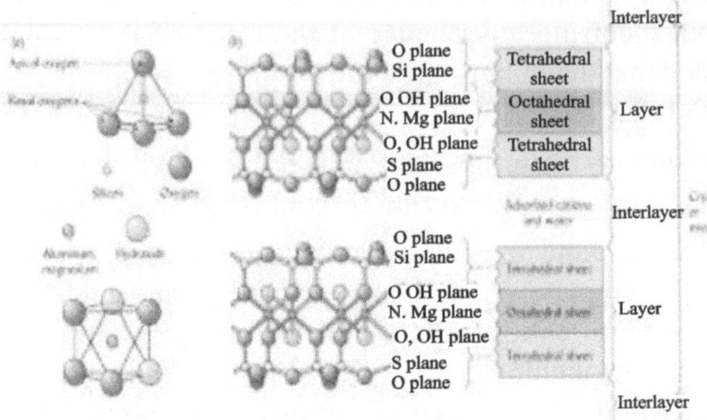

Figure 5.4. Structure of 2:1 type silicate

a) **Smectite group** (Expanding 2:1 type silicate): The minerals in this group is also called montmorillonites (Figure 5.5). The variable basal spacing can accommodate more than two layers of water molecules with no clearly defined upper limit. The most dominant dioctahedral members under the smectite groups are Montmorillonite, Beidilite and Nontronite. The half cell structural formula of Montmorillonite is; $0.33M^{+}xH_2O\ (Al_{1.67},Mg_{0.33})Si_4O_{10}(OH)_2$. Smectitic minerals are mostly found in Vertisols, Mollisols, Alfisols and to some extend Entisols.

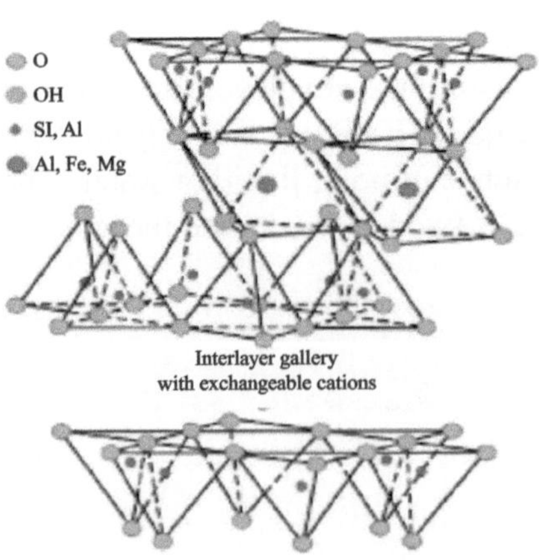

Figure 5.5. Structure of Montmorillonite

b) **Hydrous mica group** (Non-expanding 2:1 type silicate): The Non-expanding 2:1 type (Illite) mineral (Figure 5.6) is characterised with the inclusion of K^+ in the interlayer position in 12 fold coordination in the hexagonal cavity existing between the oxygen planes of tetrahedral sheets of two adjacent layers. Because of the strongly bound tetrahedral sheet of two adjacent layers mediated by the K^+ ions, the expansion of this mineral is restricted on wetting and behaves more like Kaolinite than Smectite in terms of adsorption of water , cations and plasticity of the clay. The characteristics basal spacing of 10A is found in Illite by the XRD and the mineral is dominant in Mollisols , Alfisols , Spodosols, Inceptisols and Entisols.

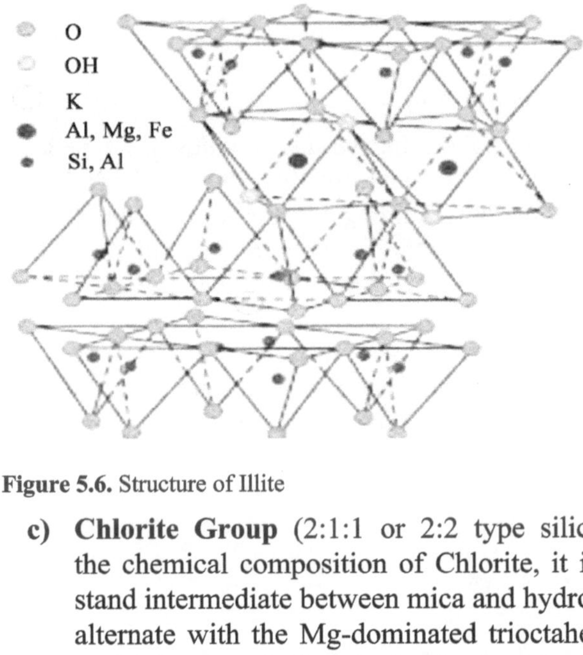

Figure 5.6. Structure of Illite

c) **Chlorite Group** (2:1:1 or 2:2 type silicate clay): While analysing the chemical composition of Chlorite, it is observed that the mineral stand intermediate between mica and hydrous oxides in which the mica alternate with the Mg-dominated trioctahedral [Mg $(OH)_6$] or brucite or Al-dominated dioctahedral hydroxide $Al_2(OH)_6$ in regular sequence (Figure 5.7). The Chlorite in general is having 2:1:1 or 2:2 structure for occupying the octahedral position by iron and magnesium rather than aluminium. The brucite sheet is developed with positive charge by substitution of Al^{+3} for Mg^{+2}. The chlorite is non-expanding type of mineral having the specific surface area 70-150 m^2/g and spacing of 14A^0. This mineral is found in Alfisols, Mollisols and Aridisols.

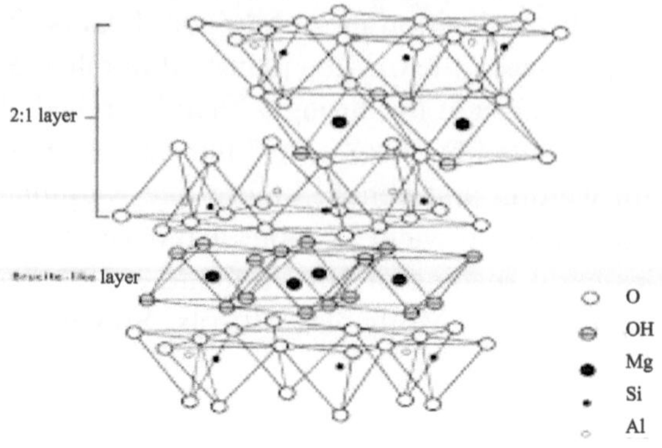

Figure 5.7. Structure of Chlorite

4. Palygroskite group (2:1 ribbon type)

Palygroskite and Sepiolite contain ribbon like structure with a 2:1 type layer structure, each ribbon being linked to the next inverted SiO_4 tetrahedral sheet along with a set of Si—O bonds. Thus, creating tetrahedral point in opposite directions in adjacent ribbons. The structure of Palygroskite (Figure 5.8) is based on amphibole or double chains $[Si_4O_{11}]$ which run parallel to the axis. An open channel of fixed dimension parallel to the chain is created for the alternation in the silicon tetrahedral sheet enforcing the corresponding alteration in the octahedral strip. One half of water molecules lie adjacent to the octahedral strips to satisfy the coordination requirement of Mg or Al and the rest takes up definite positions in the channel.

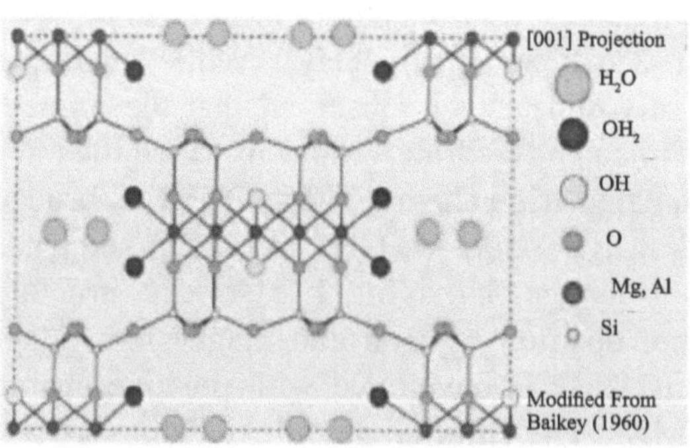

Figure 5.8. Structure of Palygroskite

5.1.2.2 Non-crystalline Silicate Clays (Allophane and Imogolite)

The members of this group do not have a definite feature under the XRD and crystallinity is not detected. The non-crystalline clays like allophane and imogolite are found in volcanic ash. Allophanes are hydrous aluminosilicates with Si/Al ratio of 2:1 to 1:1. The presume formula is $2SiO_2 .Al_2O_3.2H_2O$.

5.1.2.3 Iron and Aluminium Oxides

The examples of iron and aluminium oxides are the Gibbsite (Al-oxide) and Goethite (Fe-oxide). Gibbsite is an aluminium hydroxide mineral of the oxides and hydroxides group, with structural formula [Al(OH)$_3$]. Aluminium is a trivalent cation for which, each of the hydroxyls is bonded to only two aluminiums, and thus, only two-thirds of the available octahedral sites are occupied. Aluminium hydroxide being the amphoteric in nature, has both basic and acidic properties and is closely related are aluminium oxide -hydroxide, [AlO(OH)] and or alumina (Al_2O_3). These compounds together are the major components of the aluminium ore, bauxite. The XRD on iron and aluminium oxides give strong diffractive peaks at 4.18A^0 and 4.82A^0 for Goethite and Gibbsite respectively. The DTA of goethite, gibbsite and show strong endothermic peak between 290-350^0 C.

5.1.2.4 Mixed or Interstratified Layer Clay

In this clay minerals, different kinds of clay layers are arranged alternately with each other. The mixing or interstratification in vertical stacking can be regular (ordered), segregated regular, or random. Specific groups of clay minerals do not occur independently of one another. In a given soil, several clay minerals are formed in an intimate mixture. Besides, some mineral colloids have properties and composition intermediate between those of any two minerals. Such minerals are termed mixed layer or interstratified (Figure 5.9) minerals where the individual layers within a given crystal may be of more than one type, e.g. chlorite-vermiculite and mica - smectite are used as mixed- layer minerals.

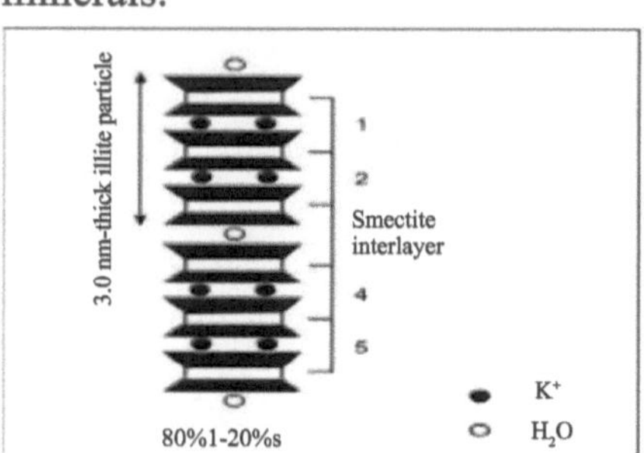

Illite-Smectite mixed layer

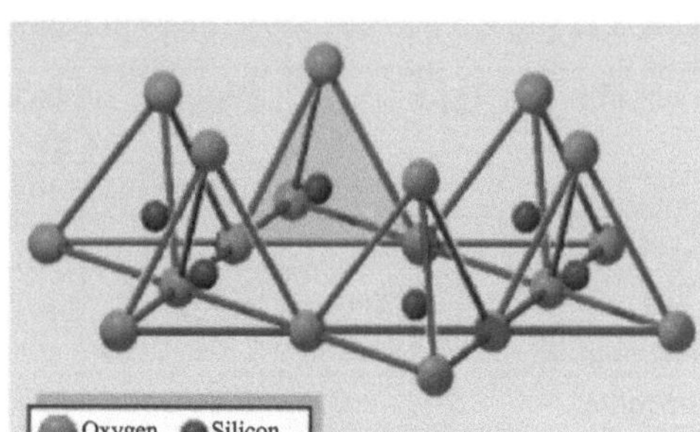

Interstratified clay

Figure 5.9. Mixed and interstratified layer clay

5.1.2.5 Organic (humus) Colloids

Organic colloids (Figure 5.10) or humus are highly coiled, dark brown long chained and polycondensed structure. Humus is the decomposed product of plant and animal residues and are composed of carbon, hydrogen, oxygen and nitrogen, instead of silicon, aluminium and oxygen, as found in clay colloids. The higher adsorptive properties for water and cation exchange capacity than the inorganic colloids is the unique property of organic colloids. Humus is amorphous in nature with variable sizes. Humus has three components *viz.*, fulvic acid, humic acid and humin. The humus colloid exhibits the pH-dependent charge. The strongly bound hydrogen is not easily replaced by other cations humus from the humus resulting in a low negative charge on the surface.

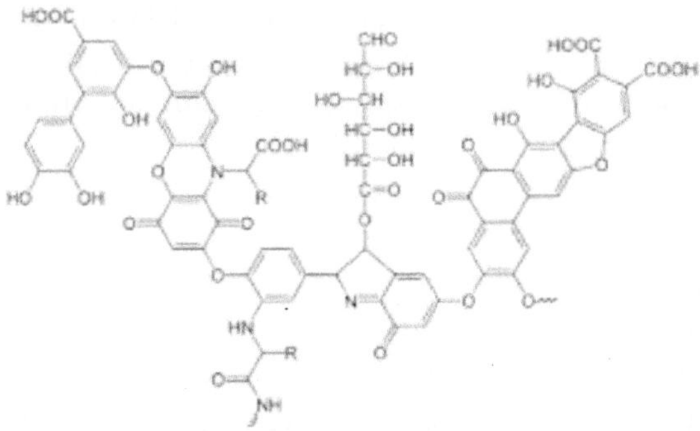

Figure 5.10. Organic colloids

The major properties of clays (Table 5.1) could identify the different nature and characteristics of the clay colloids. The type, specific surface area, c-axis spacing and ion-exchange capacity could demarcate the adsorption capacity for ions and expanding /non-expanding properties of the clay minerals.

Table 5.1. Major properties of selected soil colloids

Colloid	**Type**	**Specific surface area (m^2/g)**		**c-axis spacing (A^0)**	**Ion exchange capacity (cmol/kg)**
		External	**Internal**		
Kaolinite	*1:1 silicate*	5-30		7.2	-1 to -15
Illite	*2:1 silicate*	70-175		10.0	-10 to -40
Vermiculite	*2:1 silicate*	70-120	600-700	10-15	-100 to -200
Smectite	*2:1 silicate*	80-150	550-650	10-20	-80 to -150
Chlorite	*2:2 silicate*	70-100		14.1	-10 to -40
Humas	*Organic*	Variable		-	-100 to 500

Hysteresis

The phenomenon of dependence of soil moisture characteristics curves on soil moisture history is called hysteresis (Figure 5.11). If the temperature and solute concentrations remain constant, the soil moisture content (θ) at any matric potential (Ψm)is lower when the soil is wetting (sorption) than drying (desorption). The soil water relationship is governed by the alternate wetting and drying (swelling and shrinkage) of clay. The hysteresis is affected by the non-uniformity of soil pores. Some pores may be filled with water while some microspores may be filled with air restricting the entry of water in those pores. The water from the pores will be drained at low matric potential.

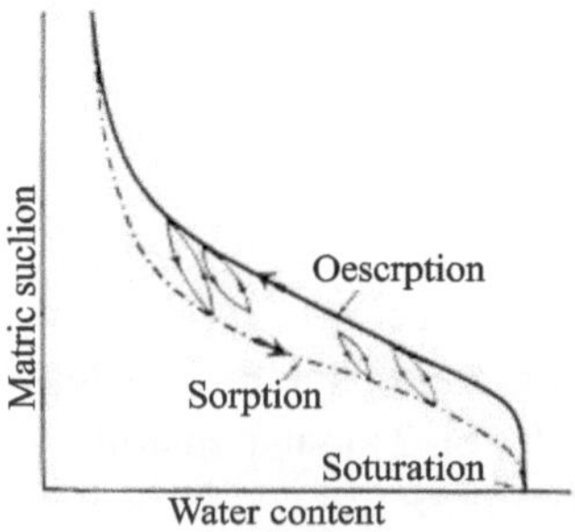

Figure 5.11. Hysteresis effect on soil

5.2 Identification of Clays

There are different methods of identification of clays. Some of the important methods of identification of clays are appended below:

5.2.1 Methods based on crystal structure

a) X-Ray diffraction analysis

b) Electron diffraction analysis

X-ray diffraction analysis

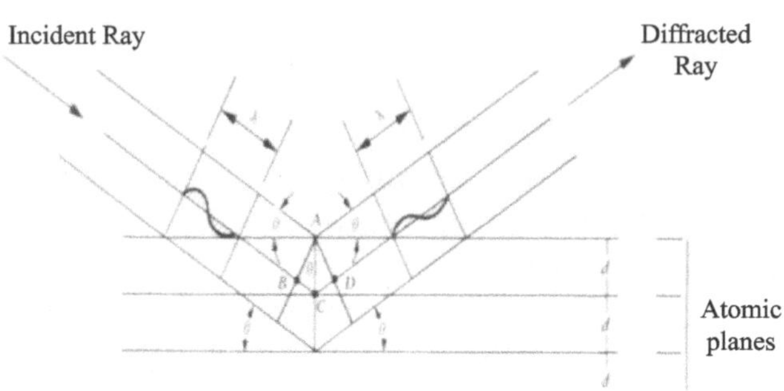

Figure 5.12. X-ray diffraction pattern on clay sample.

The distance of atomic planes d (Figure 5.12) can be determined based on the Bragg's equation.

$$n\gamma = 2d\sin\theta$$

$$\text{or, } d = n\gamma/2\sin\theta$$

where, n is an integer and γ is the wavelength (1-2A^0)

Different clays minerals have various basal spacing (atomic planes). For example, the basal spacing of kaolinite is 7.2 Å. The XRD pattern (Figure 5.13) could identify the c-axis spacing of the individual clay.

Chemical and Thermal Pre-treatment are required before the analysis and following Diffractograms are produced.

1. Mg-saturation of clay
2. Mg-saturated and glycerol solvated
3. K-saturated and heated to 300°C
4. K-saturated and heated to 550°C

The flowchart for sample prepation are given below;

- Soil/Clay → dispersed in water (Sodium hexametaphosphate) → less than 0.2 microns (centrifugation) → 2 ml of 1% or 2% suspension → slide preparation → dry

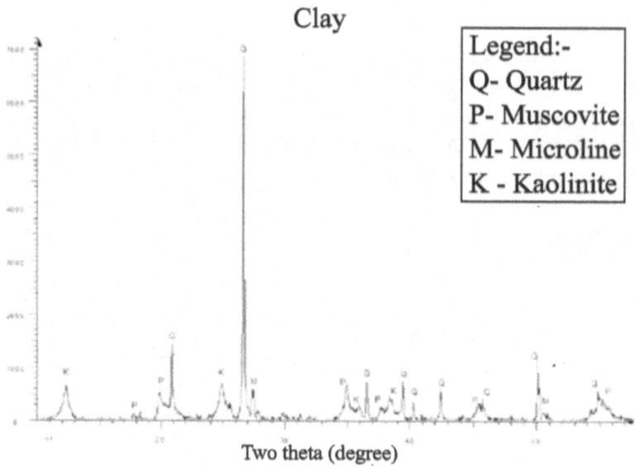

Figure 5.13. The XRD pattern of clay

5.2.2 Methods Based on Thermal Properties

Here temperature plays the key role for determining the clay mineral identification. Some important methods are given below:

a) Differential Thermal Analysis (DTA)
b) Thermogravimetric Analysis (TGA)
c) Differential Thermogravimetric Analysis (DTGA)

5.2.2.1 Differential Thermal Analysis (DTA)

Differential thermal analysis (DTA) consists of simultaneously heating a test sample and a thermally inert substance at constant rate (usually about 10 °C/ min) to over 1000 °C and continuously measuring differences in temperature and the inert material ΔT. Endothermic (take up heat) or exothermic (liberate heat) reactions can take place at different heating temperatures. The mineral types can be characterized based on those signatures shown in the figure 5.14.

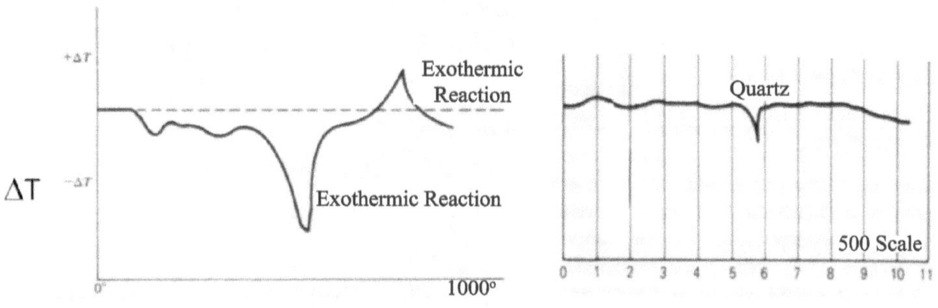

Figure 5.14. Exothermic and Endothermic peak

5.2.2.2 Thermogravimetric Analysis (TGA)

In TGA (Figure 5.15), the mass of a sample monitored as a function of temperature, is one of the oldest analytical techniques used in clay mineralogy. Since then, TGA has been used to obtain a variety of information on minerals, particularly hydrous phases, such as clays and zeolites.

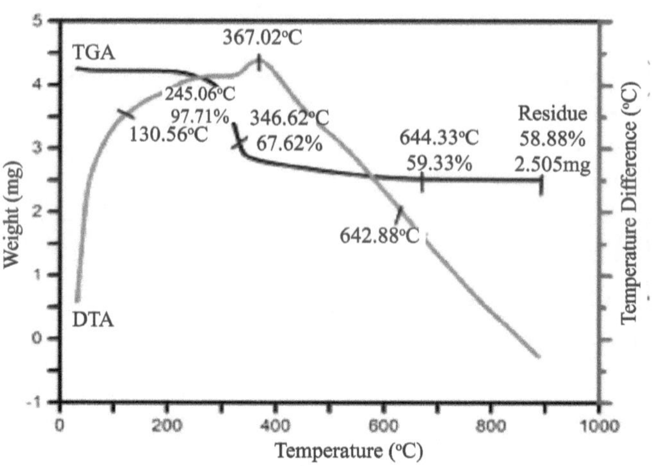

Figure 5.15. Thermogravimetric peaks of clay

5.2.2.3 Differential Thermogravimetric Analysis (DTGA)

When a sample is heated, reactions will occur in general. If the heat absorbed or emitted during these reactions is compared to the characteristics displayed by an inert material (usually anhydrous alumina) undergoing the same temperature change, a trace is obtained during differential thermal response against temperature. If the sample absorbs heat with respect to the standard, a trough occurs in the trace (an endothermic reaction). Where the sample evolves heat compared to the standard, a peak occurs (an exothermic reaction). A heating rate of 10°C/min is usually adopted up to 1000°C. Some important changes are held during the thermal reactions on clay which are as follows:

a) **Dehydration (Endothermic)**
   i) adsorbed water (driven off between 100-300°C)
   ii) crystal lattice water [the (OH) ions] - termed dehydroxylation which causes a complete destruction of the mineral structure.

b) **Exothermic Reactions**
   i) Crystallisation: new crystals may form from amorphous materials, or from old crystals destroyed at lower temperatures,
   ii) Oxidation: e.g. organic matter between 250-450°C.

The endothermic peak under DTA (Table 5.2) is used to identify the different clay minerals.

Table 5.2. DTA for common soil minerals

Mineral	**DTA Peak (endothermic)°C**
Kaolinite	500-550(Symmetrical)
Illite	550-650
Montmorillonite	850-900
Vermiculite	700-850
Chlorite	750-800
Halloysite	480-530 (Asymmetrical)

Source: Raman and Ghosh (1974)

5.2.3 Methods based on Physico-Chemical and Chemical Properties

a) Electrometric and Viscometric analysis
b) Surface area
c) Cation Exchange Capacity
d) Cation Exchange Capacity delta value (CEC Δ)
e) Elemental chemical analysis
f) Y-index of Martin and Russel

5.2.3.1 Electrometric and Viscometric Analysis

Identification of clay by Electrometric analysis is governed by pH-metric titration curves (Figure 5.16) of the respective clay minerals and or measuring the apparent viscosity for the clay minerals.

a) pH-metric titration curve
b) Viscosity neutralization curve

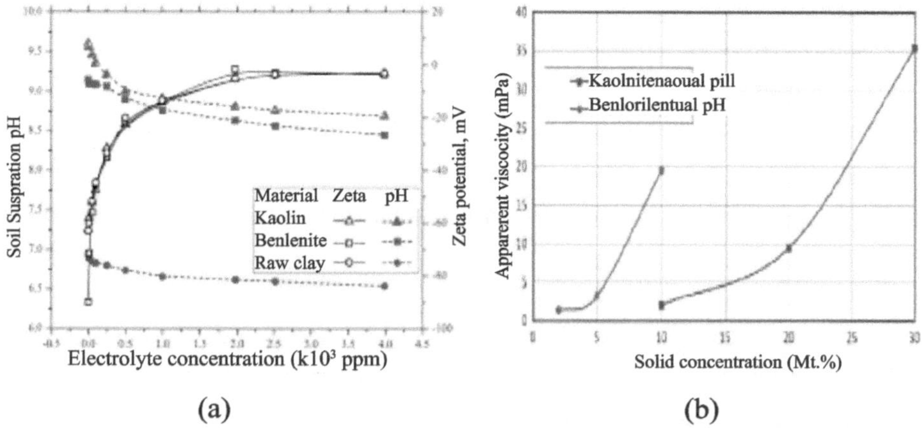

Figure 5.16. Potentiometric (a) and Viscometric (b) curves of clays

5.2.3.2 Specific Surface Area

The assumption underlying the determination of specific surface is that, when exposed to an environment of glycol, the clay layer will adsorb a monomolecular layer of the glycol. The specific surface area of a clay is the amount of dominant mineral present in clay fractions. Clay layers will adsorb polar molecules from the liquid as well as from the gaseous phase. Either ethylene glycol (EG) or glycerol is used as the environment. The test procedure involves determining the dry mass of sample, exposure in a glycol atmosphere and determination of (Table 5.1) mass of sample plus glycol retained.

$$\text{Specific Surface } (m^2/g) = Mg/\,(Ms \times 0.00031)$$

Mg = mass of glycol retained

Ms = mass of dried sample

0.00031 = mass of glycol (in grams) required to form a monomolecular layer on a surface area of $1m^2$

5.2.3.3 Cation Exchange Capacity (CEC)

Cation exchange capacity (CEC) measures those ions bound loosely to the mineral surface by charge deficiency in the particular crystal lattice. One procedure used to determine CEC involves pre-heating the sample to remove soluble salts and organic matter and then saturating potential sites with $CaCl_2$. Sodium acetate is used to replace the Ca^{2+} which is then determined in the washing solution by titration.

5.2.3.4 Cation Exchange Capacity - Delta Value [CEC Δ]

The CEC Δ value is determined by the following flow chart to identify clay minerals.

Soil sample treated with H_2O_2 →Fe-oxide extracted and treated with 2% Na_2CO_3 solution (1hr.)

Soil sample treated with NaOAc (pH 3.5) at boiling for 15 mints.

Both samples are washed with 1N neutral NaOAc →CEC determined ....X_1

Again the samples are Ca-saturated with 1N neutral CaOAc →CEC determined .....X_2

Difference $(X_1\text{-}X_2)$ = CEC Δ

The endothermic peak under DTA (Table 5.2) is used to identify the different clay minerals.

5.2.3.5 Elemental Chemical Analysis

The nature of the dominant clay minerals in soils are ascertained through elemental chemical analysis. The total elemental composition of the samples in terms of having Si^{+4}, Al^{+3}, Fe^{+3}, Ti^{+4}, Mg^{+2}, Ca^{+2}, Na^{+1}, Mn (with different valance), K^{+1}, phosphate, water etc. between 105°C and 1000°Cor below 105°C. The amount of mica present can be calculated from the K_2O content which is for Muscovite is about 10.6% and Biotite is 9.4%. Similarly, the content of MgO is the indicative of chloride on vermiculite. The SiO_2/R_2O_3 ($R_2O_3 = Al_2O_3$ or Fe_2O_3) when more than 2.0 is indicative of montmorillonite dominating clay and if less than 2.0 is meant for Kaolinite dominating soil.

5.2.3.6 Y-index of Martin and Russel

The presence of dominant clay minerals in soils by combining the following three factors was assessed by Martin and Russel (1952).

Three indices

1. Ethylene glycol retention
2. Total K_2O content
3. CEC of clays

Y-index is calculated as follows:

$$\mathbf{Y = 1/3\ [Ax10/6 + Bx10/20 + Cx10/25]}$$

where,

A= % of K_2O

B= Internal retention of EG in mg/g

C= CEC of the material

0-5 = Kaolinite like

5-25 = Illite like

≥25 = Montmorrilonite like

5.2.4 Methods based on shape and size/morphology

Electron microscopy of clays/Scanning electron microscopy

5.2.4.1 Electron Diffraction Analysis

The Principal of identification of clay minerals is same as XRD. Here, fast and high energy electron beam (deBrogile) of wavelengths of 0.05^0A is used which are defracted by the identical lattice planes. The exposure time is only the few seconds. However, because of the absorption, the electron beam has low penetrating energy which could give information on few nanometer from the surface of the crystal.

e.g. Identification: Attapalgite – the characteristic (110^0) reflection with an interplanar spacing of 10.5^0A.

5.2.4.2 Electron Microscopy of clay

The particles with less than 0.2μm can not be seen under the ordinary microscope but can be visible with radiation having small wave length compared to the visible light. The wavelength of electron beam can be made small under high voltage, the electron get accelerated and resolution can be made better. The magnification may be as high as 200,000 times. When the specimen is exposed to the electron beam , the beam strikes on the fluorescent screen and enlighten the screen in contrast with the specimen as a dark object. The electron microscopy gives sufficient information regarding morphology, chemical composition and structure of the individual specimen.

5.2.5 Methods Based on Other Properties

a) Infrared Spectral (IR) analysis
b) Staining test

5.2.5.1 Infrared (IR) spectral analysis

The IR may be used for both the amorphous and crystalline clay samples when XRD is not sufficient to identify clay minerals. The major functional groups of the clay may be identified through this methods. Band positions in IR analysis are indicative of the unit of frequencies or wave number which is related to the wavelength.

$$\upsilon = 10^4/\chi$$

where, υ= frequency or wave number, *i.e.* number of waves per cm (cm^{-1})

χ = wave length (μm)

The IR spectral analysis is done in the region of 4000 to 600 cm^{-1} wave number. The higher wave number (4000-3000 cm^{-1}) is correlated with the vibrations of absorbed water molecules or octahedral OH-groups, while the lower wave number (1400-800 cm^{-1}) is due to the vibration of Al-OH or Si-O. Different clay minerals exhibit different wave bands (Figure 5.17) with clay film samples.

Flow chart

Clay (1.5mg) of < 2 micron size + IR quality KBr→clear disc preparation in vacuum (pressure 60tons/sq inch) placed in IR radiation → absorption spectra

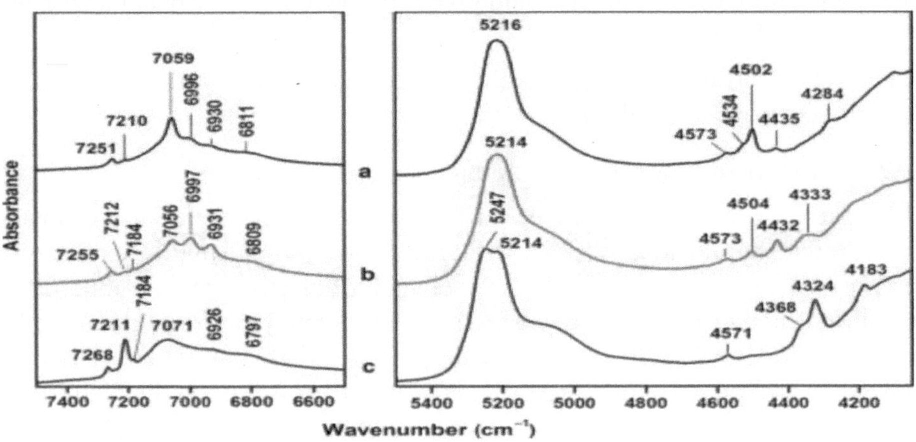

Figure 5.17. Infrared (IR) - Spectral curves

5.2.5.2 Staining Test

The test can be applied to the relatively pure clay minerals and to soil or clays free from iron oxide and other non-crystalline constituents. On adsorption of organic dyes by clays could yield semiquantitative estimate of the amount of clay present in a rock, and staining thin sections with some dyes (e.g. methylene blue or benzidine) might reveal the clay-component and the clay content of the rock is counted under the microscope. A cyclic aromatic hydrocarbon, like Benzidine undergoes a redox reaction with clay minerals whereby it is changed to a monovalent cation with the mechanism of the benzidine-blue reaction (Solomon *et al.*,1968), where, benzidine is oxidised at clay surfaces or edge sites to form a semiquinone which is blue (Theng, 1981).

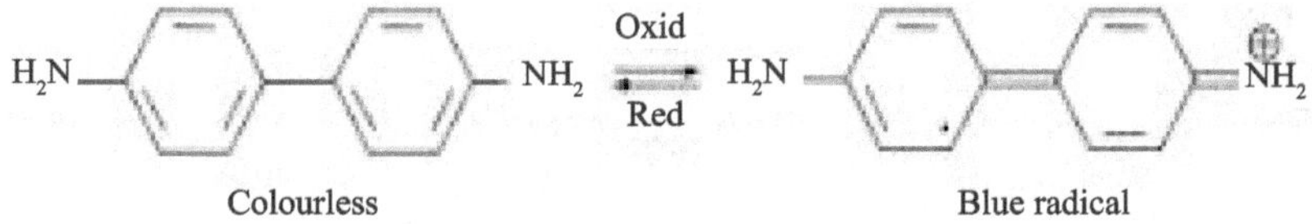

The electron transfer is believed to occur between the colourless benzidine and some of the ferric ions at planar surfaces or aluminium at crystal edges or both. The oxidation of iron bearing smectite ensuring the presence of ferric iron could enhance the colour reaction (Solomon *et al.* 1968). Besides, it has been observed that, the benzidine-blue reaction is dependent upon the pH of the aqueous solution in which it takes place (Solomon *et al.* 1968; Furukawa and Brindley, 1973).

5.3 Sources of Charges on Silicate Minerals

There are two types of charges commonly found on clay surfaces.

Constant Charge (structural) – Through the process of Isomorphous substitution among the ions,– both negative and positive charges can be developed. This is also called the permanent charge.

Variable Charge – This is also called the pH-dependent charge. With the addition of acid or alkali the charge characteristics of the clay will differ (Figure 5.18). The negative charges are associated with silicate clay minerals by the following ways :

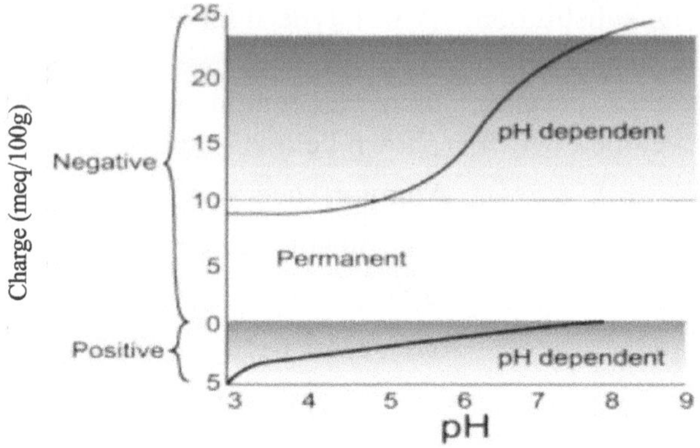

Figure 5.18. Sources of charges on clay

a) **Dissociation of OH ion at the exposed crystal edges** The unsatisfied valences at the broken edges of silica and alumina sheets may evolve dissociation of OH ion at the exposed crystal edges. The external surfaces of mineral having the exposed oxygen and hydroxyl groups may act as exchange sites (Figure 5.19). These groups are attached to Si and Al atoms within their respective sheets. At high pH, the H^+ of these OH^- radicals dissociates slightly and colloidal surface is left with a –ve charge carried by the oxygen. The loosely held H^+ is readily exchangeable to react with other ions. The presence of such group in the clay would result a negative charge. These are the pH- dependent charges of inorganic colloids. The magnitude of pH - dependent charges varies with the type of colloids.

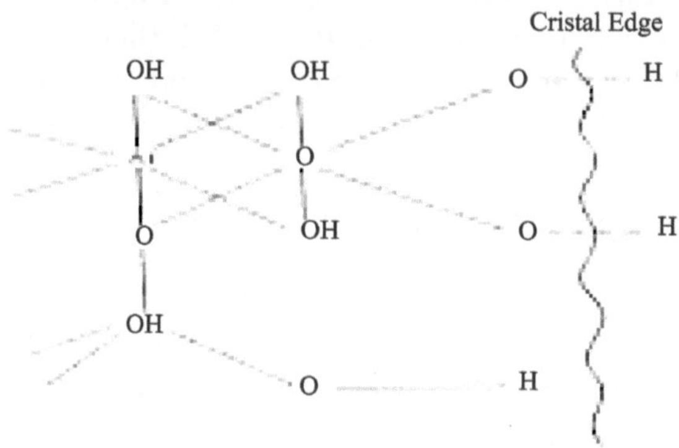

Figure 5.19. Development of negative charge at the exposed crystal edges

b) **Ionic substitution (Isomorphous substitution)** The overall –ve charge carried out by clay crystal is the substitution of one atom by other. The ions having similar size in a crystal structure of the minerals is known as ionic or isomorphous substitution. In 2: 1 type minerals when the aluminium atoms have been substituted by the magnesium atoms an unsatisfied negative valence (because a three valent atom is replaced by a two valent one) is developed which is shown below:

Alumina sheet	Alumina sheet
(No substitution)	(Mg substituted for Al)
$O^{--}Al^{+++}OH^{-}$	$O^{--}Mg^{++}OH^{-}$
No net charge	one excess negative charge

Similarly, when substitution of three-valent atom (such as Al) for one of the four-valent silicon atoms in silica sheet leaves an unsatisfied negative valance. This represents below.

Silica sheet	Silica sheet
(No substitution)	(Al substituted for Si)
$O^{--}Si^{++++}O^{--}$	$O^{--}Al^{+++}O^{--}$
No net charge	one excess negative charge

The charges developed from ionic substitution are not pH dependent. So they are referred to as permanent charge (Figure 5.20). There is marked differences between the CEC and AEC in Kaolinite for almost equal distribution of positive and negative charge. In Kaolinite, pH-dependent is the only source, while in montmorillonite negative charge is mainly due to ionic substitution and only ¼ by dissociation of OH ion and as a result, a wide difference between CEC and AEC in the montmorillonite is observed. In the silicate clay minerals, the negative charges developed due to isomorphous substitution and dissociation of OH^- ion can attract positively charged cations like Ca^{+2}, Mg^{2+}, K^+, Na^+, H^+. NH_4^+ which are adsorbed on the clay surface as exchangeable and available forms to plants.

The cations and water are adsorbed on the clay surfaces. In general the cations adsorbed on the surface of silicate clays are in the order of:

$$Al^{+3} > Ca^{+2} >= Mg^{2+} > K^+ = NH_4^+ > Na^+$$

Certain cations are especially prominent under natural conditions.

i) Humid region soils $Al^{+3} > H^+ > Ca^{+2} = Mg^{2+} > K^+ > Na^+$

ii) Well drained and semi arid soils $Ca^{+2} > Mg^{2+} > Na^+ > K^+ > H^+$

iii) Sodic or Alkali soils $Na^+ > Ca^{+2} > Mg^{2+} > K^+ > H^+$

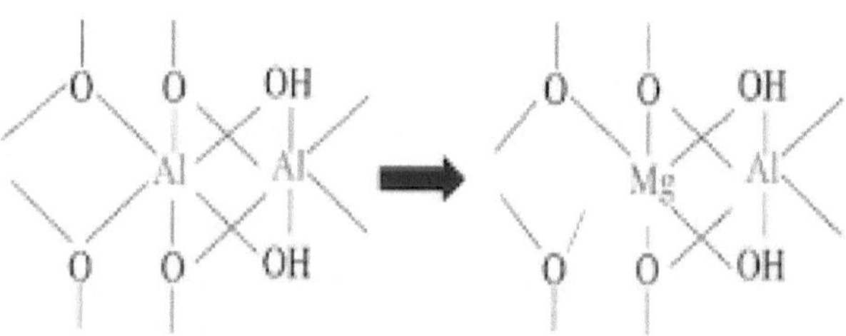

Figure 5.20. Development of permanent charge

c) Due to dissociation of carboxylic/phenolic group from the humus, negative charge on clay may be developed (Figure 5.21).

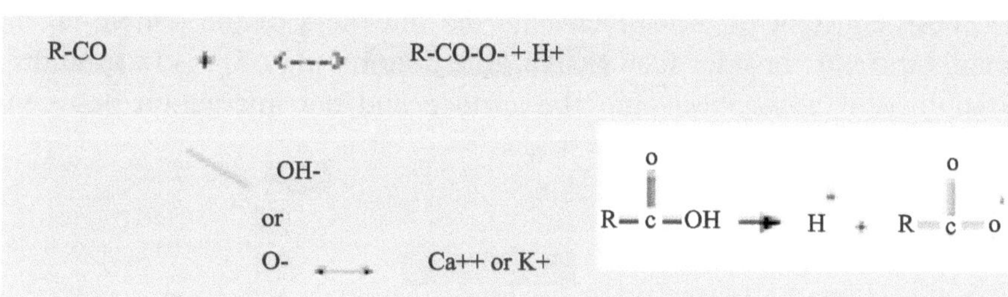

Figure 5.21. Development of negative charge on organic matter

5.4 Electrical Double layer

A double layer (also called an electrical double layer, EDL) is a structure that appears on the surface of an object which might be a solid particle, a gas bubble, a liquid droplet, or a porous body. The Double Layer (DL) refers to two parallel plates/layers of charge surrounding the object. The first layer, the surface charge (either positive or negative), consists of ions adsorbed onto the object due to chemical interactions while the second layer is composed of ions attracted to the surface charge via the Coulombic force and is loosely associated with the object (Figure 5.22). The free ions that move in the fluid under the influence of electrostatic attraction and thermal motion rather than being firmly adsorbed on the surface is called the "diffuse layer". The two layers thus can constitute the double layer also known as **diffuse double layer** (DDL).

5.4.1 Development of double layer

The double layers are formed due to the fixed and diffuse part of the ions in the clay. The distribution of the ions in the clay matrix varies for the change in distance from the fixed and diffuse part of the array or may be for the different ionic concentrations. Some of the empirical models on DDL are given below.

5.4.1.1 Helmholtz plane

When a solid or liquid ionic conductor (electrolyte), is in contact with each other a common boundary (interface) among the two phases appears. Hermann von Helmholtz realized that charged electrodes immersed in electrolyte solutions repel the co-ions of the charge while attracting counter-ions to their surfaces. Two layers of opposite polarity can be formed at the interface between electrode and electrolyte. It was observed that, an electrical double layer (DL) is essentially a molecular dielectric and stores charge electrostatically. Below the electrolyte's decomposition voltage, the stored charge is linearly dependent on the voltage applied. This model predicted a constant differential capacitance independent from the charge density which depends on the dielectric constant of the solvent and the thickness of the double-layer. This model did not consider factors like diffusion/mixing of ions in solution, the possibility of adsorption onto the surface and the interaction between solvent dipole moments and the electrode.

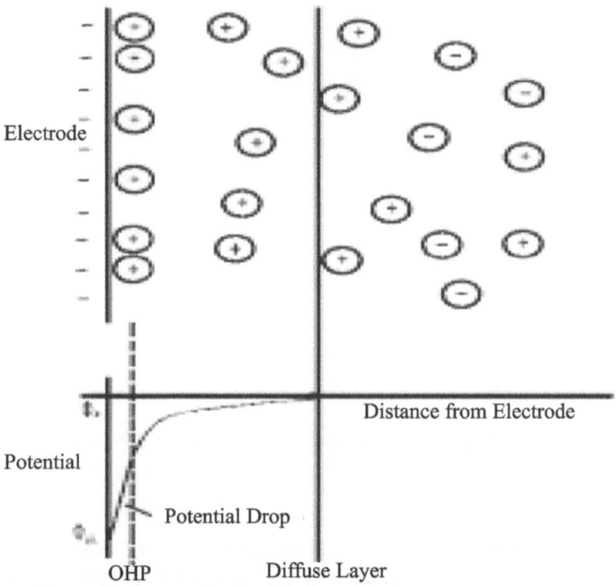

Figure 5.22. Electrical Double Layer

5.4.1.2 Gouy–Chapman layer

Louis Georges Gouy (1910) and David Leonard Chapman (1913) observed that, capacitance was not a constant and that it depended on the applied potential and the ionic concentration. The "Gouy–Chapman model" introduced a diffuse model of the DL. In this model the charge distribution of ions as a function of distance from the metal surface allows Maxwell–Boltzmann statistics for application. Thus, the electric potential decreases exponentially away from the surface of the bulk solution.

5.4.1.3 Stern Layer

The limitations of the Gouy-Chapman model at the highly charged DLs, made thinking to combine the Helmholtz model with the Gouy-Chapman model by Otto Stern (1924). In Stern's model, some ions adhere to the electrode as suggested by Helmholtz, giving an internal Stern layer, while some coming from a Gouy-Chapman diffuse layer (Figure 5.23).

The ions' finite size and consequently an ion's closest approach to the electrode in accordance with the ionic radius is considered in Stern layer. The Stern model has its own limitations, namely that, it effectively treats ions as point charges, assumes all significant interactions (Coulombic) in the diffuse layer and assumes dielectric permittivity to be constant throughout the double layer and that viscosity of fluid is constant in the plane.

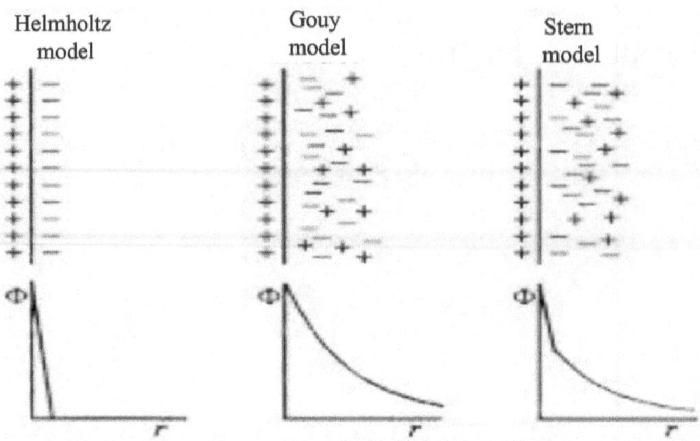

Figure 5.23. Schematic representation of ion and potential distribution in the double layers

5.4.1.4 Triple Layer

While understanding the theory of DDL, a concept of triple layer model was introduced by Yates *et al.* (1974). The layer of adsorbed potential determining ions (Figure 5.24) was considered as an integral part of the solid clay surface (Kleijn and Oster,1983). This layer which is considered to be the first layer is located in the liquid phase and the other layer adjacent to the 1st layer ,located also in the liquid phase is divided into two sublayers. They are called Inner Helmholtz Layers (IHL) and Outer Helmholtz Layers (OHL). The IHL adsorbed the traditional counter ions whereas the OHL the changing diffuse layers. The triple layer will not develop in absence of potential –determining ions.

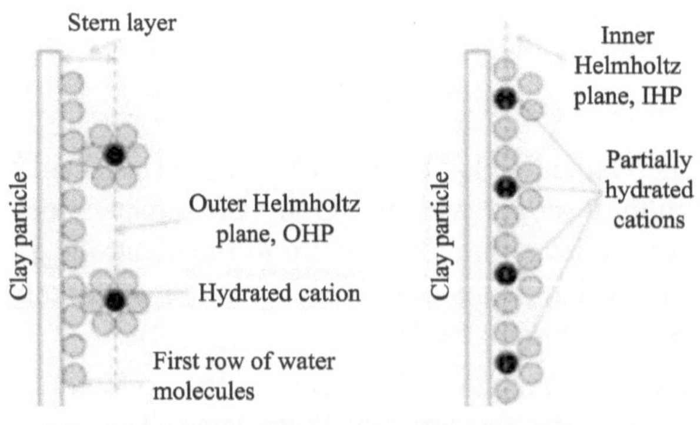

Figure 5.24. Triple layer formation on clay colloids

5.4.2 Important properties of DDL

i) The thickness of the diffuse double layer decreases as the electrolyte concentration in the water increases. An equal equivalent concentrations of electrolyte, divalent cations decrease the thickness of the double layer more strongly than that of monovalent cations and trivalent ions more strongly than divalent.

ii) The cation concentration gradient in this diffuse double layer sets up an electrical potential difference between the surface of the clay particle and the bulk of the solution, hence work done to transfer a cation from the surface of the particle to the bulk of the solution can be assessed.

iii) The proportion of cations that dissociate into Gouy diffuse layer can be estimated by measuring the activity of the cations around the clay particles using a specific membrane electrodes.

iv) The concentration of the anions in the solution is lower in the diffuse double layer than in the bulk solution . There is possibilities of negative absorption for which difficulty arises while measuring the adsorption of ions by soils. When the concentration of the adsorbate (substance adsorbed on adsorbent) is less on the surface of the adsorbent (substance on which another substance is adsorbed) than in the bulk, then it is known as Negative adsorption.

5.4.2.1 Factors Affecting Diffuse Double Layer

The diffuse double layers are affected by the following factors:

i) **Concentration of Total Electrolyte** With an increase in total electrolyte concentration, the thickness of the diffuse double layer decreases.

ii) **Valence of the Counter Ion** The higher the valence of the opposite or counter ion, the thinner or smaller will be the thickness of the diffuse double layer.

iii) **Dielectric Constant** With an increase in the dielectric constant of the medium, the thickness of the diffuse double layer increases.

5.4.2.2 Limitations of Diffuse Double Layer

i) The main limitation of the structure of diffuse double layer (DDL) is to state a distinct line between the micellar (inner) and the inter-micellar (outer) solutions.

ii) The structure and properties of diffuse double layer is applicable only to the large flat or plane surfaces of the clay minerals. The pattern of distribution of ions is characterised by the charge densities at the edges and corners and cavities of the clay minerals, assuming the same pattern of distribution of ions for a flat or plane surface of clay minerals.

iii) The theories of the diffuse double layer are not valid for the soil organic matter (SOM) partly due to structure of humus and partly due to forces other than coulombic resulting separate kind of bonding among the adsorbed ions.

iv) The Gouy theory is valid only for colloids having a low surface-charge density and at a very low concentrations of the solution when the ionic dimension is considered as negligible. The thickness of Gouy's diffuse double layer decreases with increasing valence and concentration of the electrolyte in solution.

5.4.2.3 Zeta Potential (ζ)

Zeta potential (ζ) is also termed as electrokinetic potential in colloidal dispersions and is expressed as volts (V) or millivolts (mV). The zeta potential is the electrical potential in the interfacial double layer (DL) at the location of the slipping plane relative to a point in the bulk fluid away from the interface. In other words, zeta potential is the potential difference between the dispersion medium and the stationary layer of fluid attached to the dispersed particle.

The zetapotential is caused by the net electrical charge contained within the region bounded by the slipping plane, and also depends on the location of that plane and is widely used for determination of the magnitude of the charge. However, zeta potential is not equal to the Stern potential or electric surface potential in the double layer because these are defined at different locations. The magnitude of the zeta potential could indicate of colloidal dispersions, the degree of electrostatic repulsion between adjacent and charged particles in a dispersion. For the small molecules/particles a high zeta potential will confer stability, i.e., the solution or dispersion will resist aggregation, but when the potential is small, attractive forces may exceed this repulsion and the dispersion may break and flocculate. So, colloids with high zeta potential (negative or positive) are electrically stabilized while colloids with low zeta potentials tend to coagulate or flocculate in the system. Zetapotential can be expressed as ;

$$\zeta = 4\pi\sigma d/D$$

where, ζ = Zeta potential or, Electrokinetic potential; d= Distance, D= Dielectric constant,

σ = Surface charge density

5.5 Adsorption of ions in soil

Adsorption is a surface process of accumulation of a gas or liquid on a liquid or solid surfaces. Furthermore, the strength of the interaction between the adsorbent (the substrate onto which chemicals attach) and the adsorbed molecules are also important.

Physisorption — Van der Waals interactions between substrate and adsorbate (the molecule that is adsorbed).

Chemisorption — Chemical bonds involved (covalent bonds usually) in sticking the adsorbate to the adsorbent.

More energy is involved in Chemisorption than physisorption. The binding energy of interaction is important to make differences between the two processes.

Sorption is a physical and chemical processes by which one substance becomes attached to another one. The sorption of a particular ions is specific and is governed by physical and chemical parameters.

Absorption – It is the incorporation of a substance in one state into another of a different state (e.g., liquids being absorbed by a solid or gases being absorbed by a liquid). The nutrients taken up by the plants are catered by the absorption phenomena. The process through which ions or molecules are removed from solution and accumulate within existing solid constituents or at the solid-liquid interface.

- Adsorption – The physical adherence or bonding of ions and molecules onto the surface of another phase (e.g., reagents adsorbed to a solid catalyst surface).
- Ion exchange – An exchange of ions between two electrolytes or between an electrolyte solution and a complex. The exchange process of two solids are possible when they are in close contact with each other.

The reverse of sorption phenomena is desorption .

5.5.1 Adsorption Isotherm

Definition

Adsorption Isotherm is a curve that expresses the variation in the amount of gas adsorbed by the adsorbent with pressure at constant temperature. The variation of the amount of ion adsorbed at different concentration of a solution at a particular temperature can constitute the adsorption isotherm.

5.5.1.1 Freundlich Adsorption Isotherm

Freundlich's empirical equation between the amount of gas adsorbed by a unit mass of solid adsorbent and pressure at a particular temperature is expressed using the following equation –

$$x/m = k.P^{1/n}\ (n > 1) \quad (5.1)$$

where 'x' is the mass of the gas adsorbed on mass 'm' of the adsorbent at pressure 'P', while 'k' and 'n' are constants that depend on the nature of the adsorbent and the gas at a particular temperature.

The mass of the gas adsorbed per gram of the adsorbent is plotted against pressure in the form of a curve to show the relationship. Here, at a fixed pressure, physical adsorption decreases with increase in temperature. The curves reach saturation at high pressure. Linearizing the equation 4.1 and taking log on both sides of the equation ,

$$\log x/m = \log k + 1/n \log P \qquad (5.2)$$

The plot log x/m against log P shows (Figure 5.25) a straight line for validation of the Freundlich isotherm otherwise, it is not. The slope of the straight line gives the value of 1/n, while the intercept on the y-axis gives the value of log k.

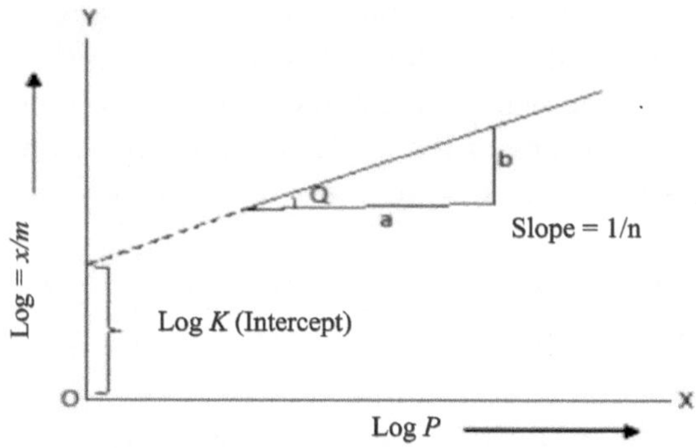

Figure 5.25. Freundlich Adsorption Isotherm for Phosphorus

Limitations of Freundlich Isotherm

Freundlich isotherm explains the behaviour of adsorption processes of ions on a specific medium. The value of 1/n can be between 0 and 1, therefore, the equation holds good only over a limited range o pressure.

- When, $1/n = 0$, x/m is constant, the adsorption is independent of pressure.
- When $1/n = 1$, $x/m = kP$, i.e., $x/m \propto P$, adsorption is directly proportional to pressure.

Experimental results support both of the above mentioned conditions. At high pressure, the experimental isotherms always seem to approach saturation and fails at higher pressure.

Step-K and Constant rate K (CR^K)

Haylock (1956) coined the terms Step-K and CR^K as a measure of potassium reserve in soil. Step-K is determined by subtracting the CR^K from the amount of potassium released in each step of successive extraction and summing up all the values obtained after initial removal of water soluble – K and exchangeable-K (Maclean, 1961).

5.5.1.2 Langmuir Adsorption Isotherm:

The Langmuir adsorption isotherm is a semi-empirical isotherm derived from a proposed kinetic mechanism. This isotherm was based on the assumptions that, dynamic equilibrium exists between adsorbed gaseous molecules and the free gaseous molecules.

For adsorption of ions or solute as a monolayer , adsorption is expressed as

$$x/m = k_1k_2C/(1+k_2C) \quad (5.3)$$

where, x is the amount of ions adsorbed on amount of clay (m), C is the concentration of the equilibrium solution and k_1 and k_2 are constants.

On linearizing the equation 4.3, the equation becomes

$$1/x/m = 1/\ k_1k_2C + 1/k_1 \quad (5.4)$$

$$\text{or, } C/x/m = 1/k_1k_2 + C/k_1 \quad (5.5)$$

The equation 5.5 states that, x/m becomes constant at high concentration. In other words, at high value of C, the surface of the adsorbent becomes saturated and adsorption of ions reaches a maximum. There are in general four assumptions on the Langmuir adsorption isotherm.

1. There is uniform surface of the adsorbent that is, all the adsorption sites are equivalent.
2. There should not be interaction among the Adsorbed molecules .
3. All adsorption occurs through the same mechanism.
4. At the maximum adsorption, only a monolayer is formed: molecules of adsorbate do not deposit on other, already adsorbed.

Langmuir suggested that adsorption takes place through this mechanism:

$$A(g) + B(s) \underset{\text{Deaorption}}{\overset{\text{Adsorption}}{\rightleftharpoons}} AB$$

where,

A(g) = unadsorbed gaseous molecule

B(s) = unoccupied metal surface

AB = Adsorbed gaseous molecule.

The direct and inverse rate constants are k and k^{-1}

5.5.1.3 BET adsorption isotherm

Brunauer-Emmett-Teller (BET) theory explains the physical adsorption of gas molecules on a solid surface and serves as the basis for an important analytical technique for the measurement of the specific surface area of materials. The BET theory (1938) applies to systems of multilayer adsorption and usually utilizes probing gases that do not chemically react with material surfaces as adsorbates to quantify specific surface area. The probing adsorbates along with N_2 are utilized, even at the lower frequency, allowing the measurement of surface area at different temperatures and measurement scales. These have included argon, carbon dioxide, and water. Specific surface area is a scale-dependent property and no single true value of specific surface area is definable, and thus quantities of specific surface area determined through BET theory may depend on the adsorbate molecule utilized and its adsorption cross section. The concept of the theory is an extension of the Langmuir theory, which is a theory for monolayer molecular adsorption, to multilayer adsorption with the following hypotheses:

1. gas molecules physically adsorb on a solid in layers infinitely.
2. gas molecules only interact with adjacent layers .
3. the Langmuir theory can be applied to each layer.
4. the enthalpy of adsorption for the first layer is constant and greater than the second.
5. the enthalpy of adsorption for the second or higher layers is the same as the enthalpy of liquefaction.

The resulting BET equation is

$$\frac{P}{V_{total}(P-P_0)} = \frac{1}{V_{mono}C} + \frac{c-1}{V_{mono}C}\left(\frac{P}{P_0}\right)$$

$$V_{total} = \frac{V_{mono}C\left(\frac{P}{P_0}\right)}{\left(1-\frac{P}{P_0}\right)\left(1+C\left(\frac{P}{P_0}\right)-\frac{P}{P_0}\right)} \tag{5.6}$$

Where, V is the volume of gas adsorbed at pressure P, Vm is the volume of gas required to form monolayer over the entire surface of the adsorbant, P_0 is the gaspressure required for monolayer saturation at the temperature of the experiment, C is the constant for the particular gas,adsorbent and temperature. The surface area of soil can be calculated using the BET equation by multiplying the value of V with cross sectional area of the adsorbate molecules like water,ammonia etc.

5.6 Electrometric properties of soil colloids

5.6.1 Potentiometric and Conductometric titrations

A freshly prepared acid clay is a colloidal acid having the anion of colloidal dimension while the exchange positions are occupied by H^+ ions. However, the lattice Al^{3+} ions are partially mobilized laterally by the exchangeable H^+

ions. As a consequence, the H-clay behaves as a mixed H-Al system. The pure H-clay exhibits the characteristics of a strong acid but the mixed sample behaves like a moderately weak to weak acid. The pH-potentiometric curve (Figure 5.26) of the freshly prepared acid clays (Montmorillonite or Kaolinite) shows the signs changing one from that of a strong acid to one characteristics of weak polybasic acid on aging. Thus, the number of inflection points in the titration curve increases, each one corresponding top different acidic groups such as polymeric Al-acidic and silanol acidic group.

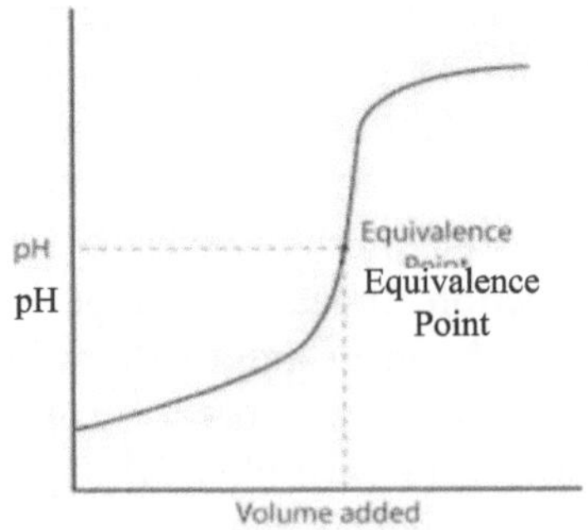

Figure 5. Volume Added n curve

On conductometric titration (Figure 5.27) of such acid clay, similar changeover from the characteristic shape of a strong acid to that of a weak polybasic acid are also taken place on aging. Here also two to three breaks are observed.

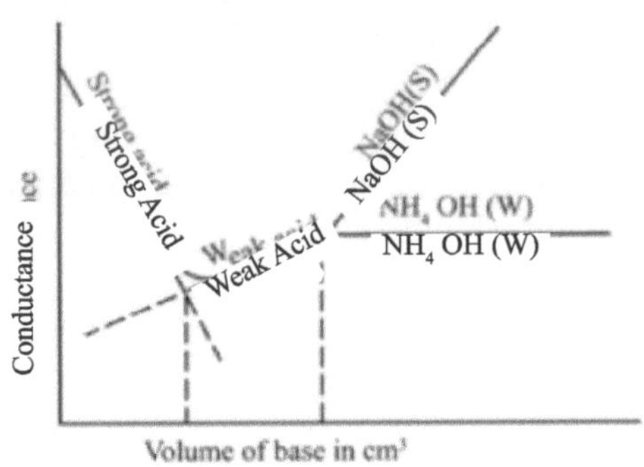

Figure 5. Volume of Base in cm^3 itration curve

When the freshly prepared acid clay is kept for several weeks it is called aged clay. The lattice Al^{3+} ions are partially mobilized laterally by the exchangeable H^+ ions. As a consequence, the H-clay behaves as a mixed H-Al system.

The conductometric titration curve of such acid clays on aging reveals similar change from the characteristic shape of a strong acid to that of a weak-polybasic acid. Here also two to three breaks are observed.

5.6.2 Spectroscopic Studies of Humic Substances

Soil - humic substances are highly coiled polycondensed in nature having dark brown colour, high molecular weight and resemble characteristic visible spectrophotometric and infra red (IR) spectroscopic traits. Among the humic and fulvic fractions of soil organic matter, the humic substances are higher polymers having darker colours at the red end of the visible spectrum. The dark colour of humic substances is the direct consequence of a greater degree of extensive delocalization of Π electrons in humic substances (HA) compared to the corresponding fulvic fractions (FA). Such behaviour is reflected in the relative absorbance (OD) for radiations of HA and FA at 465 nm and 665 nm (denoted by E_4 and E_6 respectively). The HA fraction being darker (towards red end) in shade, would naturally absorb light more at 665 nm than that at 465 nm. On the other hand, the value (E_4/E_6) is expected to be higher on FA as compared to that of the corresponding HA fraction. Indeed a change in the ratio accompanying by a change in pH of the light absorbing medium in soil in a coiled (folded) state to minimize their contact surface area for water for thermodynamic restrictions partially unfolded with raise of pH of the soil solution under thermodynamic compulsion (Sanyal,1984) This alters the E_4/E_6 ratio depending upon the predominance of aliphatic /aromatic character of the FA/HA substances.

The positive (if) values of $\Delta(E_4/E_6)$ signifies the unfolding of the hidden aromatic parts inside the coiled HA/FA molecules at higher pH, in higher proportion than the corresponding aliphatic parts. This ordinarily lead to a lowering of (E_4/E_6) value resulting this positive value of $T(E_4/E_6)$.

Despite the complexity of natural organic molecules, **IR spectroscopy** is useful in selectively studying carboxyl groups due to their high polarity (i.e., high **infrared** activity) and pH behaviour (Figure 5.28).

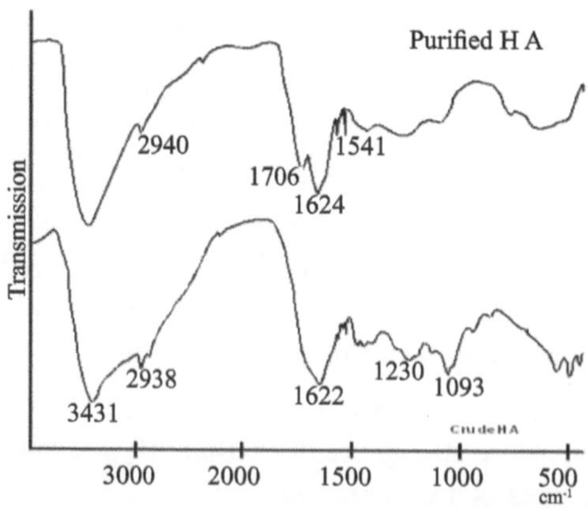

Figure 5.28. FT-IR spectra of crude and purified humic acids derived from peat soil

5.7 Point of Zero charge or Zero Point Charge on Clay

The pH at which the net surface charge of the soil colloids become zero is called the zero point charge (ZPC) or point of zero charge (PZC). At soil pH greater than ZPC, the mineral is negatively charged and exposed attractions for cations while at a soil pH less than ZPC, the mineral surface will develop positive charge attracting the anions. The ZPC of hydrous oxides of iron and aluminium can be shown (Figure 5.29) as:

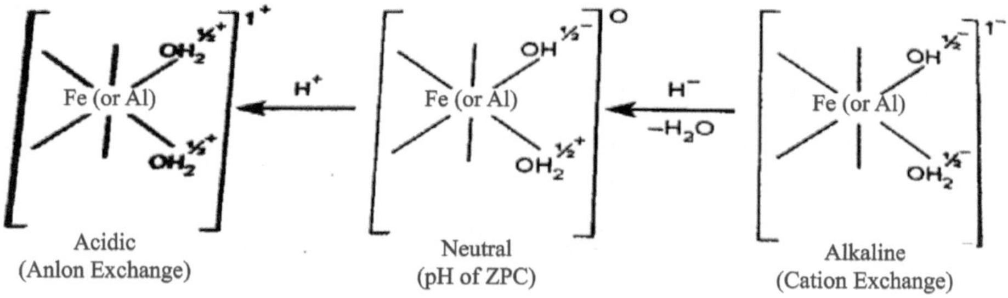

Figure 5.29. Development of ZPC on iron or aluminium

The small amount of positive charge may be developed at the broken edge of the clay due to lowering of pH (Figure 5.30)

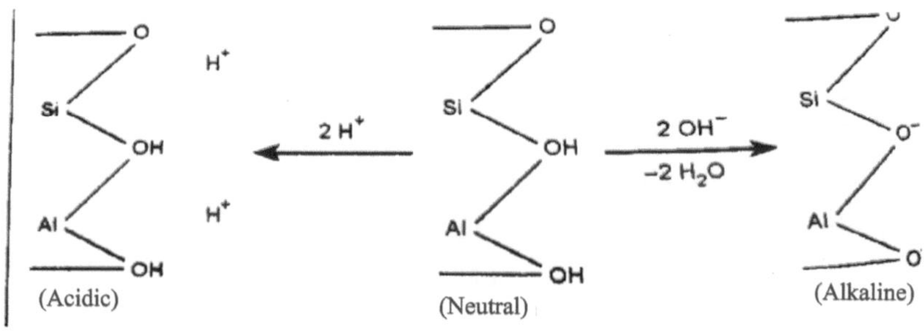

(Adapted from Sanyal, S. K. 2002)

Figure 5.30. Development of charge on clay surfaces at different pH

5.8 Surface charge characteristics of soils

Point of Zero Charge(PZC) or Zero Point Charge (ZPC) is associated with the variable charge characteristics of clays where the ZPC is the particular pH where the net surface charge is zero. It requires to estimate the net permanent charge of the soil to understand the position of ZPC.

It has been observed that, fixation of ions, particularly the phosphate , sulphate, borate etc. is controlled by the variable charge components of the soils which intern will tend to predict the P-availability to plants and P-fixation in soils. The shift of ZPC is governed by the surface charge parameters of the soils.

From the equation (Sposito,1984),

$\sigma_t = \sigma_P + \sigma_\psi$ (5.7)

where, σ_t = net surface charge density

σ_P = permanent charge density

σ_ψ = variable surface charge density

again,

ψ_0 = $(RT/F)[\ln(H^+)/(H^+)_{ZPC}]$ (5.8)

= 0.059 (ZPC – pH) at 25° C

again, $\sigma_P = [2C_\varepsilon RT/\pi]^{1/2} \sinh [ZF\, \psi_0/2RT]$

$\sigma_{v=} [2C_\varepsilon RT/\pi]^{1/2} \sinh 1.15Z(ZPC\text{-}pH)$ (5.9)

where,

σ_t = Total net surface charge density ($Coulomb.m^{-2}$)

σ_P = Permanent surface charge density ($Coulomb.m^{-2}$)

σ_v = Variable surface charge density ($Coulomb.m^{-2}$)

C = Ionic concentration in the bulk soil solution ($mol.m^{-3}$)

ε = Static permittivity of aqueous medium = $\varepsilon_0 D$

ε_0 = Permittivity in vacuum ($8.85 \times 10^{-12} F.m^{-1}$)

D = Dielectric constant of the medium

Z = Valence of the counter ion

Ψ_0 = Surface potential (V)

R = Universal gas constant (8.317J/molK)

F = Faraday Constant (96487 Coulomb/mol)

T = Absolute temperature (K)

(H^+) = Hydrogen ion activity in soil solution.

$(H^+)_{ZPC}$ = Hydrogen ion activity at ZPC.

Thus shifting of ZPC towards the lower or higher pH zone depends on the presence of organic matter, degree of weathering on soils, leaching of the cations from the soil surface and the available charge density on clays. The surface potential (Ψ_0) of a clay may be more negative when there is higher H^+ ion concentration in the adjacent soil solution and the ZPC will be at lower pH level. On the other hand, accumulation of oxides/hydroxides of Fe and Al in soil could shift its ZPC towards higher pH. Hence, corresponding charge characteristics of clays may predict the processes of soil development (Sanyal, 2002).

5.9 Specific and Non-specific Sorption

The specific adsorption is considered as the positive adsorption and is a type of covalent bonding of solutes by the sorbents. This kind of adsorption is also called chemisorption (Schwertmann and Taylor, 1977). When the adsorption process is catered by the electrostatic force of attraction between the solutes and clay is known as non-specific adsorption. The specific adsorption to complexation of solutes is inner sphere surfaces of clay minerals. (Sposito 1989, 2008) and the non-specific adsorption for the complexation of solutes is the outer sphere surfaces of clay minerals.

Questions

Objective

1. Element essential for deflocculation of soil particles is
   a) Potassium b) Calcium
   c) Magnesium d) Sodium
2. Y-index was given by
   a) Martin and Russel
   b) Ramamoorthy & Subramanian
   c) Larsen
   d) Schofield
3. What is the commonly used soil:water ratio for pH-determination
   a) 1:1 b) 1:2
   c) 1:2.5 d) 1:5
4. Of the following , which is a 2:2 types of clay minerals
   a) Chlorite b) Polygroskite
   c) Sepiolite d) All
5. Which of the following is a soil dispersing agent ?
   a) Sodium fluoride b) Sodium chloride
   c) Sodium phosphate d) Sodium hexametaphosphate
6. The reason for soil flocculation is
   a) Lowering of zeta potential b) Increase of zeta potential
   c) No effect of zeta potential d) None is true
7. Which of the following is a non-silicate mineral?
   a) Kaolinite b) Calcite
   c) Amphibole d) Vermiculite

8. The unit of CEC in soil is
   a) $Cmol(p+)kg^{-1}$ b) mg/g
   c) m^2/g d) cm^2/g
9. Most dominant element in Earth crust is
   a) O_2 b) Ca^{++}
   c) Al^{3+} d) Si^{4+}
10. C-axis spacing of illite is
   a) 10 Å b) 7 Å
   c) 14 Å d) 18 Å
11. Hysteresis is related with
   a) Sorption curve b) Desorption curve
   c) Sorption-desorption curve d) None
12. Metahalloysite is
   a) 2:1 type of clay mineral b) 2:2 type of clay mineral
   c) 1:1 type of clay mineral d) None of the above
13. Lime potential is
   a) pH-1/2pCa b) pH+1/2pCa
   c) pOH-1/2pCa d) pOH +1/2pCa
14. Which of the following is a 2:2 clay mineral?
   a) Montmorillonite b) Chlorite
   c) Kaolinite d) Vermiculite
15. In plastic consistency:
   a) Clods form b) Soil puddles
   c) Soil sticks d) Soil adheres to plough
16. Differential thermal peaks of montmorillonite clay is
   a) 500-550 b) 850-900
   c) 750-800 d) 550-650

Subjective

1. Define soil colloids.
2. What are the types of soil colloids ?
3. What are the C-axis spacing of 1:1; 2:1 (expanding and non-expanding); 2:2 types of soil colloids.
4. What are the different methods of identification of clay based on physico-chemical and chemical properties ?

5. What is pH-independent charge on clay ?
6. What is triple layer in soil ?
7. Outline the differences between sorption and desorption of ions in soil.
8. What is potentiometric curve ?
9. Outline the concept of E_4/E_6 of soil humic fractions.
10. What is the dilution effect on diffuse double layers ?

Descriptive

1. Outline in brief the properties of soil colloids.
2. Briefly describe the 2:1 type of clay.
3. How does charge brucite layer develop in 2:2 type of clay ?
4. Describe the methods of identification of clay.
5. What are the factors affecting negative charge on clay ?
6. Briefly describe the theories of diffuse double layer.
7. Describe the adsorption models of ions on clay and their limitations.
8. Outline the characteristics of potentiometric and conductometric titration. of acid clay.
9. Briefly enumerate formation of clay-humus complexation.
10. Describe the factors affecting diffuse double layer and electrokinetic potential in soil.

Suggested Readings

Bear, F.E. (1964). Chemistry of the Soils. Oxford and IBH.

Brady, N.C. and Weil, R. R. (2007). The Nature and Properties of Soils. 14th Edition, Prentice Hall, USA.

Furukawa, T. and Brindley, G.W. (1973). Adsorption and oxidation of benzidine and aniline by montmorillonite and hectorite. Clays and Clay Minerals, 21: 279-288.

Greenland, D.J. and Hayes, M.H.B. (1981). Chemistry of Soil Processes. John Wiley and Sons.

Grim, R.E. (1968). Clay Mineralogy. Mc Graq Hill.

Raman, K.V. and Ghosh, S.K. (1974). Identification and quantification of minerals in clays. In: Minerology of Soil Clays and Clay Minerals, Bulletin No.9 (S.K. Mukherjee and T.D. Biswas, Eds). Indian Society of Soil Science, New Delhi, pp. 117-142.

Sanyal, S.K. (1984). Structure of water solutions of organics- Hydrophobic hydration. Chem. Edn. (UGC), 1(2): 14-18.

Solomon, O. H., Loft, G.C. and Swift, A. (1968). Reactions catalysed by minerals- IV: The mechanism of the benzidine blue reaction. Clay Minerals 7: 389-397.

Sposito, G. (1989). The chemistry of Soils. Oxford Univ. Press.

Tan, K.H. (2005). Principles of Soil Chemistry. Marcel Dekker Inc., New York.

Theng, B.K.G. (1971). Mechanisms of formation of colored clay-organic complexes. A review. Clays and Clay Minerals 19: 383-390.

6

Ion Exchange Phenomena

The exchange of ions (cations or anions) are important phenomena in soil for the growth and nutrition of the crops. The oppositely charged ions are being adsorbed by the clay colloids which are subjected to be released or exchanged reversibly from the soil solution. The participating ions in the exchange process are called exchangeable or counter ions. The activity of ions depends on the pH of the soil. Hence, factors like, soil texture, organic matter, concentration of ions could govern the ion exchange phenomena in soil.

6.0 Ion Exchange

Ion exchange is a reversible process where exchange of ions are possible between the two electrolytes or between an electrolyte solution and a complex. The ion exchange processes help in purification, separation, and decontamination of aqueous and other ion-containing solutions. The ion-exchange resins (functionalized porous or gel polymer), zeolites, montmorillonite, clay and soil humus are considered as typical ion exchangers which are either cation exchangers [exchanging positively charged ions (cations)], or anion exchangers [which can exchange negatively charged ions (anions)]. The amphoteric exchangers are able to exchange both cations and anions simultaneously. However, the concurrent exchange of cations and anions can be more efficiently performed in mixed beds, which contain a mixture of anion- and cation-exchange resins, or passing the treated solution through ion-exchange resin columns.

Ion exchange may be selective for certain ions or classes of ions, depending on their chemical structure, size of the ions and their charge. Examples of ions that can bind to ion exchangers are:

i) H^+ (proton) and OH^- (hydroxide).

ii) Polyatomic inorganic ions (SO_4^{2-} and PO_4^{3-}).

iii) Organic bases, containing the amine functional group $-NR_2H^+$.

iv) Organic acids having molecules with $-COO^-$ (carboxylic acid) functional groups.

v) Monatomic ions (Na^+, K^+, and Cl^-) having single charge.
vi) Doubly charged ions like Ca^{2+} and Mg^{2+}.
vii) Biomolecules (amino acids, peptides, proteins, etc.) subject to ionization. In the ion-exchange phenomena, both the absorption and adsorption (sorption) can take place. Besides, ion-exchange is a reversible process, and the ion exchanger can be regenerated or loaded with desirable ions by washing with an excess of these ions. The ion exchange reactions among clay colloids and ammonium sulphate (6.1) and KCl (6.2) depict the stoichiometry of the reactions.

a) Ammonium Sulphate $[(NH_4)_2SO_4]$

$$\begin{matrix} & Ca^{++}\ Mg^{++}K^{+} & \\ {}^{H^+}_{Na^+} & \boxed{\text{Clay Colloid}} & Ca^{++} + (NH_4)_2SO_4 \rightleftharpoons \\ & Ca^{++}\ Mg^{+}\ K^{+} & \end{matrix} \begin{matrix} & Ca^{++}\ Mg^{++}K^{+} & \\ {}^{H^+}_{Na^+} & \boxed{\text{Clay Colloid}} & CaSO_4 \\ & Ca^{++}\ Mg^{+}\ H^{+} & \end{matrix} \quad (6.1)$$

b) Potassium chloride

$$\begin{matrix} Ca40 \\ A120 \\ H40 \\ M20 \end{matrix} \boxed{\text{Clay Colloid}} + 7KCI \rightleftharpoons \begin{matrix} Ca38 \\ A138 \\ H39 \\ M18 \end{matrix} \boxed{\text{Clay Colloid}} + 2CaCI_2 + HCI + 2MCI \quad (6.2)$$

(N.B.: M is considered to be monovalent)

6.1 Cation Exchange Capacity (CEC)

Cation-exchange capacity is defined as the amount of positively charged ions that can be exchanged per unit mass of soil, usually measured in cmol(p^+) kg^{-1} (earlier the unit was me/100g soil). Alternately, the CEC can be defined as milliequivalent of cations exchanged in 100g soil. The negative charges arising out on soil particle surfaces, especially those of clay minerals and soil organic matter is due to the CEC. The replacement of aluminium or silicon atoms by other elements with lower charge (e.g. Al^{3+} replaced by Mg^{2+}) can give the clay structure a net negative charge. This charge does not involve deprotonation and is therefore pH-independent and called permanent charge. In addition, the edges of these sheets expose many acidic hydroxyl groups that are deprotonated to leave negative charges at the pH levels in many soils. Organic matter also makes a very significant contribution to cation exchange, due to its large number of charged functional groups (phenolic OH, -COOH groups etc.). The CEC is typically higher near the soil surface, where organic matter content is highest and declines with depth. The CEC of organic matter is pH-dependent. Cations are adsorbed on the soil surfaces by the electrostatic force of attraction between their positive charge and negative charge of the surface. The surrounding water hull of the ions can prevent direct chemical bonds with the surface. Exchangeable cations thus form part of the diffuse layer above the charged surface. The binding is relatively weak and a cation can easily be displaced from the surface by other cations from the matrix.

6.1.1 Significance of Cation Exchange Phenomena

Cation exchange is an important reaction in soil fertility for correcting soil acidity and basicity. The soil physical properties may be changed also in purifying or altering percolating waters. The plant nutrients like calcium, magnesium and potassium are supplied to plants as exchangeable ions from the soil. The exchangeable K^+ and Mg^{+2} is a major source of plant available- K. The amount of lime required to raise the pH of an acidic soil is greater as the CEC is greater. Cation exchange sites hold Ca^{+2}, Mg^{+2}, K^+, Na^+ and NH_4^+ ions and slow down their losses by leaching. Sometimes the exchangeable sites adsorb fertilizer K^+ and NH_4^+ and reduces their mobility in soils. Heavy metals (Cd^{+2}, Zn^{+2}, Ni^{+2} and Pb^{+2}) are adsorbed by the sites that might be present in wastewater. Adsorption removes them from the percolating water and cleaning the water that drains into groundwater.

Percentage Base Saturation (PBS): The extent to which the adsorption complex of a soil is saturated with exchangeable basic cations is termed as base saturation. It is expressed as a percentage of the total cation exchange capacity.

% base saturation = [Exchangeable bases(cmol/kg) / CEC(cmol/kg)] x 100

The percentage of sites occupied by the basic cations can reveal the PBS. If the percentage base saturation is 50, half of the exchange capacity is satisfied by bases, the other by hydrogen and aluminium.

6.1.2 Factors affecting Cation Exchange Capacity

i) **Soil texture:** The negatively charged clay colloids attracts positively charged cations and holds them. Therefore, the cation exchange capacity of soils increases with increase in per centage of clay content in soil.

ii) **Clay:** This soils with high CEC can retain large amounts of cations and reduce the loss of cations by leaching. Sandy soils, with low CEC, retain smaller quantities of cations and therefore cations are removed from soil by leaching.

iii) **Soil Reaction:** As the pH is raised, the hydrogen held by the organic colloids and silicate clays (Kaolinite) becomes ionized and replaceable. The net result is an increase in the negative charge on the colloids that could increase in CEC of the soil.

iv) **Soil organic matter:** High organic matter content increases CEC. The CEC of clay minerals range from 10 to 150 [cmol (p^+) kg^{-1}] and that of organic matter ranges from 200 to 400 [cmol (p^+) kg^{-1}], depending upon the mineralogical composition of the clay /organic matter.

v) **Nature of clay minerals:** The CEC and specific area of the clay minerals are in the order: smectite>fine mica>kaolinite. Hence, the CEC of a soil dominated by smectite type of clay minerals is higher than kaolinite type of soils.

6.2 Anion Exchange Phenomena

The phenomena of anion exchange is similar to that of cation exchange. Under certain conditions hydrous oxides of iron and aluminium shows evidence of having positive charges on their crystal surfaces. The positive charges of colloids are due to addition of hydrogen (H^+) in hydroxyl group (OH^-) resulted in net positive charge (OH^{2+}). This positive charge will attract anions. The capacity for holding anions increases with increase in acidity. The lower the pH of the soil the greater is the adsorption. All anions are not adsorbed at equal rate. Some anions such as $H_2PO_4^-$ are adsorbed very quickly at all pH values in the acid as well as alkaline range, Cl^- and SO_4^{--} ions are adsorbed slightly at low pH but none at neutral soil, while NO_3^- ions are not adsorbed at all. Hence, at the pH commonly prevailing in cultivated soils- nitrate (NO_3), chloride (Cl) and sulphate (SO_4) ions are easily lost by leaching. In general, the relative order of anion exchange is:

$OH^- > H_2PO_4^- > SO_4^- > NO_3^-$

$$R.OH + HCI \longrightarrow R.OH_2 + Cr$$
$$R.NH_2 + HCI \longrightarrow R.NH_3 + Cr$$

This adsorbed negatively changed anion is replaceable as under:

$$\underset{(R = AI, Fe)\quad \text{soil solution}}{R.OH^+_2\ Cr + H_2PO_4^-} \longrightarrow \underset{\text{on exchange complex}}{R.OH_2.H_2PO_4} + \underset{\text{soil solution}}{Cr} \qquad (6.3)$$

In this way, there is fixation of phosphate ion, in which OH^- ion of silicate clay mineral is substitute by $H_2PO_4^-$ ion by isomorphous substitution. The $H_2PO_4^-$ ion becomes a part of the silicate clay minerals and the phosphate fixed by this way is known as colloid bound phosphate. This phosphate is not exchangeable and available to the plant. The anions like NO_3^- and OH^- are not bound at pH above 7. The adsorbing power of anions is in the order of:

$$CI^* = NO_3^* < SO_4 < PO_4$$

* are not held by the soil colloids and are lost due to the leaching.

or $OH > H_2PO_4 > SO_4 > NO_3$ (relative order of anion exchange)

6.2.1 Importance of anion exchange in nutrient availability

The phenomenon of anion exchange is similar to that of phosphate ions and their fixation. The exchange is brought mainly by the replacement of OH ions of the clay minerals.

$$\underset{\text{Available}}{\text{OH Clay} + H_2PO_4^-} \longrightarrow \underset{\text{Unavailable}}{H_2PO_4^- \text{ Clay} + OH^-} \qquad (6.4)$$

The adsorption of phosphate ions by clay particles from soil solution reduces its availability to plants. This is also known as colloid bound phosphate fixation. The phosphate ion again becomes available when lime is applied to increase the pH of acidic soil. The OH^- ion is originated not only from silicate clay minerals but also from hydrous oxides of iron and aluminium present in the soil. The phosphates ions react with the hydrous oxides also get fixed forming insoluble hydroxyl phosphate of iron and aluminium and is called saloid bound phosphate.

$$\underset{\text{Available}}{Al(OH)_3 + H_2PO_4^-} \longrightarrow \underset{\text{Unavailable}}{Al(OH)_2.H_2PO_4} + OH^- \qquad (6.5)$$

If the reaction takes place at a low pH under strongly acid conditions, the phosphate ions are irreversibly fixed and are totally unavailable to the plants.

6.2.2 Factor Affecting Anion Exchange

The followings are some of the important factors affecting the anion exchange processes.

i) **Soil pH:** Contrary to cation exchange, the capacity for retaining anions increases with a decrease in soil pH.

ii) **Valency of Ions:** Anion exchange increases with the valence of the anion but decreases with increasing valence of the cation.

iii) **Type and Amount of Clay Colloid:** Anion exchange is much greater in soils high in 1: 1 type clay minerals and those containing hydrous oxides of iron and aluminium than that of soils containing high amounts of 2: 1 type of clay minerals. Clay minerals in the montmorillonite group of expansible layer silicates usually have anion exchange capacities of the less than 5 me/100 g. On the other hand, Kaolinites can have an anion exchange capacity as high as 43 me/100 g at an acidic equilibrium pH of 4.7.

iv) **Salt Concentration:** Anion exchange generally decreases with increasing salt concentration. Addition of salts suppresses the unequal ion distribution. In general the relative order of anion exchange is

$$OH^- > H_2PO_4^- > SO_4^{2-} > NO_3^-$$

ECEC- The Effective Cation Exchange Capacity (ECEC) is defined as the total amount of exchangeable cations, which are mostly sodium, potassium, calcium and magnesium in non-acidic soils and bases plus aluminium in acidic soils. The ECEC is determined by extracting bases with neutral normal ammonium acetate solution and the extractable aluminium by one molar potassium chloride solution. The concept of ECEC was considered to understand the role of exchangeable Al^{+3} in acid soils and also to correlate with taxonomic classification to some extend.

6.3 Ion Exchange Formulas

Law of mass action : The rate of any chemical reaction is proportional to the product of the masses of the reacting substances, with each mass raised to a power equal to the coefficient that occurs in the chemical equation.

6.3.1 Law of Mass Action and Equilibrium Constant

The law of mass action states that the velocity of a reaction at a given temperature is proportional to the product of the active masses of the reacting substances. Consider the reaction

$$A+B \rightleftharpoons C+D$$

According to the law of mass action, the rate of forward reaction (v_1) is proportional to the product of the active masses of A and B denoted $[A]$ and $[B]$

$$v_1=k_1[A][B] \tag{6.6}$$

again, the rate of reverse reaction (v_2) is

$$v_2=k_2[C][D]$$

where, k_1 and k_2 are proportionality constants. Since the equilibrium is a dynamic one, i.e., $v_1 = v_2$, then we get

$$k_1[A][B]=k_2[C][D]$$

and thus

$$k_{1/}k_2=[C][D]/[A][B]=K$$

where, K is the equilibrium constant of the reaction at constant temperature. In general for any reversible reaction

$$aA + bB \rightleftharpoons cC + dD$$

$$Kc=[C]c[D]d/[A]a[B]b \tag{6.7}$$

These concentration terms refer to the values at equilibrium and K_c is denoted as the concentration equilibrium constant.

The equilibrium constant can give information on how far the reaction has proceeded. If the reaction has proceeded almost to completion the equilibrium constant is large, while if little reaction has occurred it is small. In general three equilibrium constants may be used in reactions according to the ways in which the active masses are expressed:

i) K_c = the concentration equilibrium constant where $[A]$, $[B]$ …, etc., are given in terms of the molal concentration of the species as shown above, which is usually used for aqueous solutions.

ii) K_p = the pressure equilibrium constant where $[A]$, $[B]$ …, etc., are given in terms of the partial pressures of the species as shown above, which is used for reactions involving gaseous species

iii) K_a = the equilibrium constant expressed in terms of activities of the species can be used for gases, solids or liquids.

In a homogeneous reaction, all of the reactants and products are in the same state, i.e., in an aqueous solution. However, many reactions involve mixtures of gases, liquids, solids, etc. In this case they are defined as heterogeneous reactions.

6.3.2 Gapon Equation

For a heterovalent (Mono-Di) ion exchange system ,

$$R\text{-}nA^m + mB^n \rightleftharpoons R\text{-}mB^n + nA^m \tag{6.8}$$

where, A and B are the cations having valence m and n respectively. Considering the law of mass action at equilibrium,

$$k_{ex}=[B]^m(A)^n/[A]^n(B)^m \tag{6.9}$$

where, k_{ex} = exchange constant, *i.e,* when, Ca^{+2} is being replaced by K^+ through ion exchange process , at equilibrium we obtain,

$$R\text{-}Ca + 2K^+ \rightleftharpoons R\text{-}2K + Ca^{+2} \tag{6.10}$$

$$k_{ex}=[K]^2(Ca)^1/[Ca]^1(K)^2 \tag{6.11}$$

taking the square root on both sides , the equation 5.11 transforms into

$$[K](Ca)^{1/2}/[Ca]^{1/2}(K) = \sqrt{k_{ex}}=K_G \tag{6.12}$$

This equation is known as Gapon equation ,where, K_G is the Gapon coefficient.

6.3.3 Vanselow Equation

The concentration of the adsorbed ions are proportional to the mole fractions of total occupying ions present (Vanselaw,1932).

Thus, for a mono-divalent (K^+-Ca^{++} system) cation exchange, the equation becomes;

$$([K^+]/([K^+ + Ca^{+2}])^2(Ca^{+2)}/([Ca^{+2}]/[K^+ + Ca^{+2}])\,(K^+)^2 = k$$

Rearranging the equation 6.12 , the equation becomes

$$[K^+]^2(Ca^{+2})/[Ca^{+2)}]/([Ca^{+2}][K^+ + Ca^{+2}])\,(K^+)^2 = K_v \tag{6.13}$$

Kv = Vanselow coefficient

6.3.4 Kerr-Vanselow Equation

Based on the law of mass action ,

$N_B/N_A.m_A/m_B$ = K (Kerr) for homovalent exchange (6.14)

For heterovalent exchange, the equation proposed by Vanselow and Gapon are:

k (Vanslow) = $(N_B)^{Z}{}_{A}/(N_A)^{Z}{}_{B} \cdot (a_A)^{Z}B/({}^{a}{}_{B})^{Z}A$ (6.15)

where, Z_A and Z_B are the valences of A and B ions respectively.

for Z_A =1, Z_B =2

k (Vanslow) = $N_B/(N_A)^2 \cdot (a_A)^2/a_B$ (6.16)

6.3.4 Krisnamoorthy, Davis and Overstreet (KDO) equation

Based on the statistical thermodynamics, Krisnamoorthy *et al.* (1948) proposed the equation;

$[K^+]^2/(K^+)^2 \cdot (Ca^{+2})/[Ca^{+2}]\{[3/2Ca^{+2}]+[K^+]\}=k_{KDO}$ (6.17)

where, [ ] and ( ) denote the concentration and activity of exchangeable and free ions respectively. The k_{KDO} is similar to Kv except for the factor 1.5 in the denominator for mono – divalent exchange , which for the mono-trivalent exchange is 2, besides the power of the activity of monovalent raised to 3.

6.3.5 Jenny's Concept

The kinetic equations of cation exchange phenomena in soil was conceptualised by Gapon (1933), Jenny (1936) and Davis (1945) based on the law of mass action.

$[Na^+](K^+)/(Na^+)[K^+] = k_{eq}=k_{ex}$ (6.18)

or, $(K^+)/(Na^+)= k_{ex}\{[K^+]/[Na^+]\}$ (6.19)

where, [ ] denotes adsorbed ions and ( ) denotes free ions in solution. k_{eq} = equilibrium constant. k_{ex} = exchange constant, also called *selectivity coefficient.*

when, $k_{ex} = (K^+)/(Na^+)$, then, $[K^+]/[Na^+]=1$ (6.20)

i.e., equal amounts of Na^+ and K^+ are adsorbed in soil or clay.

6.3.6 Clay Membrane Electrodes and Ionic Activity

The ionic activity in suspension was measured by clay membrane electrodes as outlined by Marshall and his co-workers. Considering ionic activity as a function of the degree of saturation (Marshall,1949,1964) , the extend of ionization of exchangeable ions and mean free bonding energy (ΔG_{bd}), the bonding energy for a monovalent ion [e.g. monovalent cation(M^+)] may be obtained from the formula,

$\Delta GM = RT \ln C_M/a_M$ (6.21)

Where, C_M = concentration of total monovalent cations [Σ(ionised + non-ionised)]

a_M = activity of the given monovalent cation (Mv)

again, $a_M = f_M . C_M$

where, f_M = activity coefficient of the given Mv(M^+)

hence, $\Delta G_M = RT\ln C_M/ f_M C_M = -RT\ln f_M = RT\ln f_M^{-1}$ (6.22)

f_M = fraction active [$f_M = (a_M/C_M)$] as termed by Marshall

f_M is accessible for the ionic activity measurement by the given clay membrane electrodes and is capable of participating in ion-exchange equilibrium.

6.3.7 Thermodynamic Approach

In this approach the energy change of reactants at equilibrium within a chemical system must be equal to the changes of the products.

If, $\Delta Gr = \sum$Free energy of products - $\sum$ Free energy of reactants = 0

If cations A and B with corresponding valance m and n are involved in an exchange reactions and if R is the exchanger ,

then,

$$R\text{-}nA^m + mB^n \rightleftharpoons R\text{-}mB^n + nA^m$$

by employing law of mass action,

$$K_{ex} = [B]^m (A)^n/[A]^n (B)^m$$

Hence, the free energy change will be

$\Delta Gr = \Delta G_r^0 + RT \ln [B]^m (A)^n/[A]^n (B)^m = \Delta Gr^0 + RT \ln K_{ex}$ (6.23)

At equilibrium, $\Delta G_r^0 = 0$, so the equation 6.23 will be ,

$\Delta G_r^0 = -RT\ln Kex$ (6.24)

If R = 1.987 cal/k/mol and converting natural logarithm to common logarithm (base 10) by multiplying with 2.3023 ,we get,

At 25⁰C temperature , ΔG_r^0 ($Kcalmol^{-1}$) = -1.364logKex

Again, from the Debye-Huckel equation,

$-\log\gamma i = Azi^2\sqrt{I}/ (1+a_i B\sqrt{I})$ (6.25)

where, A and B are constants of the solvent at a particular temperature, Z is the valence of the ion species I; I is the ionic strength and a_i is the effective ionic diameter. At 25⁰C, the values for A and B for aqueous medium are $0.5085 L^{1/2} mol^{-1/2}$ and $0.3281\ mol^{-1/2} L^{1/2} A^{-1}$ respectively. In a very dilute solution , ionic strength becomes negligible,

$(1+a_i B\sqrt{I}) = 1$ and from the equation 6.25,

$$-\log\gamma i = AZi^2/\sqrt{I} = 0.5Zi^2\sqrt{I} \quad (6.26)$$

Hence, to assess the ionic strength of a solution , the combined effect of the activities of several electrolytes in a given solution can be estimated as ;

$$I = 0.5\ \Sigma m_i Z_i^2 \quad (6.27)$$

where, mi = moles of ion/l and Z_i = the charge of the ion.

6.4 Quantity – Intensity (Q/I) relation of Potassium K)

The Q/I relations of labile soil K relate its availability or intensity (I) to the amount (**Q**) present in soil. They are all linear over the range of activity ratios commonly applied in agricultural soils and approximate to understand ion-exchange equations.

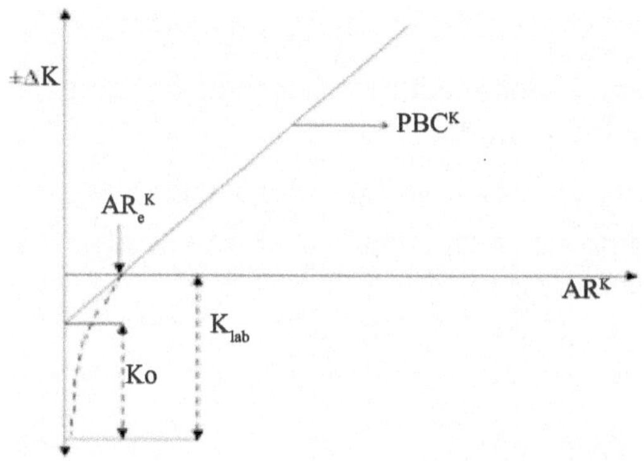

Figure 6.1. Typical Q/I relationship of Potassium

In view of the above fact, it is essential to establish the quantity relationship between exchangeable -K (Q - quantity) and the activity of potassium in the soil solution (I - intensity) to assess the availability of more labile potassium in soils to plants. The Q/I concept has been developed by Beckett which is used for predicting the status of potassium in soils. Different parameters of the above curve (Figure 6.1) have some practical implications in relation to available potassium in soils and plants.

Where, $\pm\Delta K$ = Amount through which the soil gains or loses potassium in bringing equilibrium (Q, quantity factor).

AR^K = Activity ratio of potassium (I, intensity factor).

AR_e^K = Activity ratio of potassium at equilibrium

$$AR^K = {}^aK+/\sqrt{({}^aCa^{+2} + {}^aMg^{+2})} \quad (6.28)$$

K_{lab} = Exchangeable or labile pool of potassium

K_0 = Specific sites for potassium

PBC^K = Potential buffering capacity of potassium

AR^K (Intensity Factor, I): It is calculated from the determined concentration of calcium, magnesium, potassium and sodium correcting to the appropriate activities with the help of extended Debye-Huckel theory.

AR_e^K (Activity Ratio of K at equilibrium): It is a measure of availability or intensity of labile pool of potassium in soil and can be modified by potassium fertilization. However, the availability of potassium in soils can either be increased or decreased due to application of liming materials which modifies the AR_e^K values either favourably or adversely.

K_{ex}: It is used for the estimation of labile soil potassium held in plannar (p) positions. Greater values of labile potassium, i.e., more negative ($-K_{ex}$) indicate a higher potassium release into the soil solution which results greater amount of potassium in the labile pool. However, the application of potassic fertilizers and lime in the cropped field have been found to increase the amount of potassium in the labile pool.

K_{sp} (Specific sites for potassium): It is a curved portion of the Q/I relationship while the linear portion of the curve (Q/I) is attributed to non-specific sites for potassium. Specific sites having high affinity for potassium are believed to exist on edges of clay minerals (e-positions) and in interlayer or wedge zones of weathered micas (i-positions),whereas, non-specific sites for potassium are associated with planar surfaces of clay minerals (p-positions). The i-position has the greatest specificity for K^+ which accounts for K^+ fixation in soils.

PBC^K (Potential Buffering Capacity of Potassium): The PBC^K is a measure of the ability of a soil to maintain potassium concentration in the soil solution. The PBC^K is proportional to the cation exchange capacity (CEC) of the soil that means, with an increase in CEC the value of PBC^K increases and *vice-versa* resulting from the changes in AR^K values. A high PBC^K value indicates a good potassium supplying power of soils whereas a low PBC^K value signifies very low potassium supplying power of soils indicating the need for frequent potassium fertilization. However, the higher PBC^K value may be obtained on liming which may increase in pH-dependent cation-exchange capacity. If PBC^K is low, small changes in exchangeable potassium produce large differences of potassium content in the soil solution. The PBC^K value is small in coarse textured sandy soils, where, extensive leaching, rapid plant growth etc. deplete available potassium within a few days. The relation between exchangeable and soil solution potassium is a good measure of the availability of the labile pool of potassium in soils to plants. The ability of a soil to maintain the activity ratio of K for depletion by crop uptake, leaching etc. is controlled by the nature of the labile potassium, the rate of release of fixed potassium, diffusion and transport of K^+ ions in the soil solution.

6.5 Donnan Equilibrium Concept

The Donnan equilibrium relationship is derived from the principles of thermodynamics. Under conditions of equilibrium the chemical potential μ_i of any dissolved species 'i' is the same in every phase present:

$$\mu i = \mu i^0 + RT \ln ai \qquad (6.29)$$

where, μi^0 is the chemical potential of species i in the standard state, R is the gas law constant, T is the absolute temperature, and a_i is the activity of the particular chemical species 'i' being considered. When the two electrolytes having a common ion are separated by a semipermeable membrane which is permeable to the cation and an anion of one electrolyte but is impermeable to anion of other electrolyte, then, the electrochemical potential on both sides of membrane would be equal at equilibrium, known as Donnan equilibrium.

e.g., consider a colloid in equilibrium in a bulk solution having electrolyte , say NaCl at constant temperature and pressure is characterised as

$$\mu NaCl_i = \mu NaCl_0 \qquad (6.30)$$

where, i stands for miceller solution and 0 stands for bulk solution

$$\text{hence, } \mu NaCl = \mu^0 NaCl + RT \ln (NaCl) \qquad (6.31)$$

the equation can be written as

$$\mu^0 NaCl + RT \ln (NaCl)_i = \mu^0 NaCl + RT \ln (NaCl)_0 \qquad (6.32)$$

or,

$$(NaCl)_i = (NaCl)_0 \qquad (6.33)$$

or,

$$(NaCl) = (Na^+)(Cl^-)$$

The equation 6.30 can be written with respect to ion product formula,

$$(Na^+)_i(Cl^-)_i = (Na^+)_0(Cl^-)_0$$

$$\text{or, } [Na^+][Cl^-] = (Na^+)(Cl^-) \qquad (6.34)$$

where, [ ] and ( ) denote the concentration and activity of exchangeable and free ion respectively. From the equations 6.31 and 6.32 it can be concluded that, the activity of cation and anion varies in opposite directions, so that at any point of liquid phase the activity of electrolyte is constant.

6.6 Schofield's Ratio Law

Considering the law of mass action and Gapon equation in a Na-Ca system we get,

$$[Na^+](\sqrt{Ca^{+2}})/[\sqrt{Ca^{+2}}](Na^+) = K \qquad (6.35)$$

$$\text{or, } (Na^+)/(\sqrt{Ca^{+2}}) = 1/K \,.\, [Na^+]/[\sqrt{Ca^{+2}}] \qquad (6.36)$$

If the ratio[Na^+]/[$\sqrt{Ca^{+2}}$] of adsorbed cations remain constant, the ratio of cations in solution also will remain constant. According to this law, there is no exchange of cations between the adsorbed and solution phase so long as (Na^+) /($\sqrt{Ca^{+2}}$) do not change with the changes in composition or concentration of soil solution for the Na-Ca system. As per the ratio law, on dilution, the higher valent cation in solution decreases at a faster rate compared to monovalent ions and on evaporation /drying, the concentration of higher valent cation in solution increases faster compared to monovalent ions.

6.7 Ligand Exchange

A ligand is an ion or molecule (functional group) that binds to a central metal atom to form a coordination complex. The bonding with the metal generally involves formal donation of one or more of the ligand's electron pairs. The nature of metal–ligand bonding can range from covalent to ionic and the metal–ligand bond order can range from one to three. Metals and metalloids are bound to ligands in virtually all circumstances. Ligands in a complex determined the reactivity of the central atom, including ligand substitution rates, the reactivity of the ligands themselves, and redox behaviour. Ligands are classified on the basis of charge, size (bulk), the identity of the coordinating atom(s) and the number of electrons donated to the metal . The ions derived from strong and weak acids may be of mono or poly valent in nature and are hold at a specific binding sites of clay can form a covalent bond with the surface group. In this case, there is ligand exchange in the sense that the extraneous anion acting as a ligand to the centrally located cation (in the clay lattice) substitutes an existing surface ligand (e.g., OH- ion or H_2O molecule).

6.8 Adsorption Envelope

The adsorption envelope are obtained by plotting of the corresponding adsorption maxima against the pH for each adsorbate Figure 6.2). While understanding the adsorption envelopes, it was apparent that, to get a maximum ligand exchange retention on a surface , not only the given anions, but also the corresponding undissociated acid molecules has also an equally important role in the process. The effect of pH on adsorption of phosphate at different equilibrium concentrations (Figure 6.3) may indicate the nature of sorption maxima at a given soil pH.

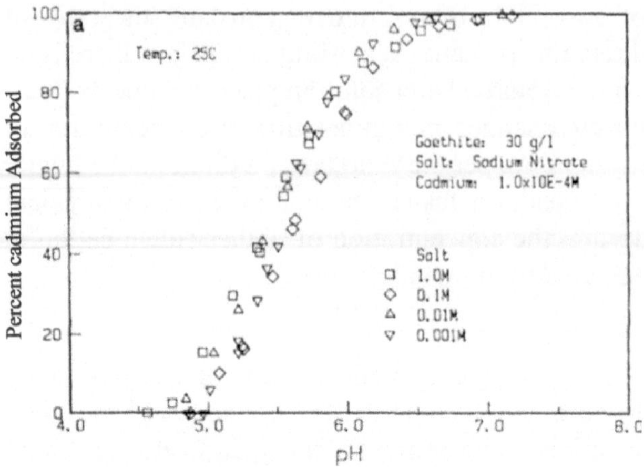

Figure 6.2. Adsorption Envelope of Cadmium

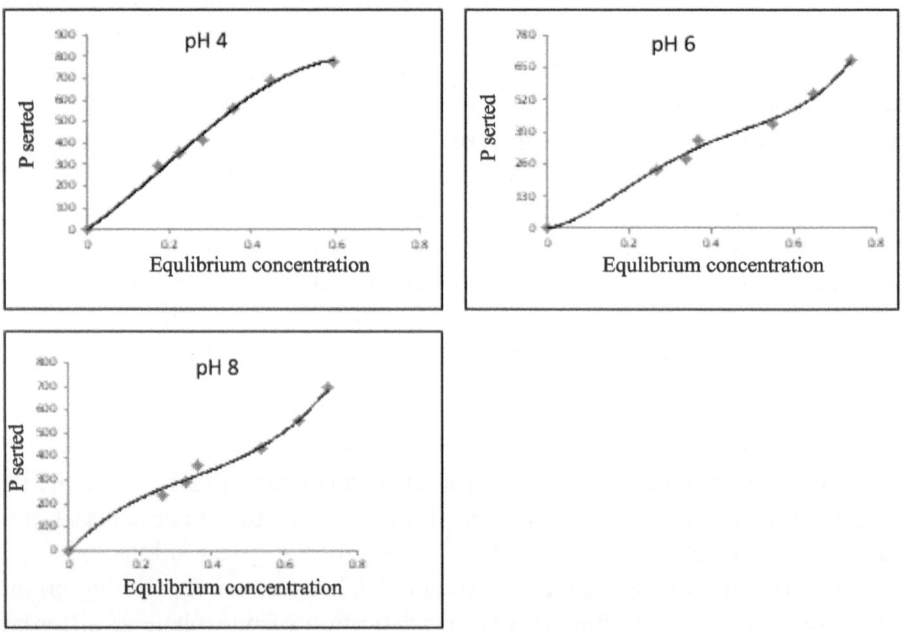

Figure 6.3. Phosphate sorption isotherms at different pHs of soil

6.9 Importance of Ion Exchange Phenomena

Ion exchange phenomena is widely used in the domestic and industrial sectors. The ion exchange process is effective in preparation of purified water for power engineering and in electronic and nuclear industries and in the water softening, water purification, water decontamination etc. Ion exchange is a method widely used in household (laundry detergents and water-filters) to produce soft water and also for the removal of nitrate from the contaminated matrix. It is widely used to separate charged molecules such as proteins. Ion-exchange processes are used to separate and purify metals. Large cation/anion ion exchangers are used in water purification of boiler-feed water. Ion exchange process can also be used to remove hardness of water by exchanging calcium and magnesium ions for sodium ions in an ion-exchange column. In pollution remediation and geotechnical engineering, ion-exchange capacity determines the swelling capacity of expanding type of clay, which can be used to adsorb pollutants. and charged ions or de-alkalinisation (removal of alkali ions from a glass surface). Regenerating waste-water through ion-exchange resin in a cyclic process is also being used.

Questions

Objective

1. At the pH range of 2-5, phosphorus is mainly fixed by
   a) Fe & Al b) Clay
   c) Ca & Mg d) Organic matter
2. The predominant ion of phosphate in soil at pH below 6.7 is
   a) $H_2PO_4^-$ b) H_3PO_4
   c) HPO_4^{2-} d) PO_4^{3-}
3. CEC of Vermiculite clay is
   a) 10-40 b) 3-15
   c) 100-150 d) 40-50
4. ZPC of silicic acid is
   a) 2.0 b) 3.0
   c) 4.0 d) 5.0
5. Fulvic acids are soluble in
   a) Acidic medium only b) Alkaline medium only
   c) All pH range d) Insoluble in all pH
6. Osmotic potential is also called as
   a) Solute potential b) Matric potential
   c) Pressure potential d) Gravitational potential
7. CEC is more in
   a) Quartz b) Kaolinite
   c) Montmorillonite d) Illite

8. Dispersion of colloidal clay is caused by

a) Ca — b) Mg
c) Fe — d) Na

9. Which ion would be readily attracted and adsorbed by soil colloids ?

a) NO_3^- — b) Ca^{++}
c) SO_4^{-2} — d) Cl^-

10. The PBC^K of soil is proportional to

a) CEC — b) AEC
c) pH — d) EC

11. Adsorption envelop in soil is related with

a) pH — b) Eh
c) pe — d) None of the above

12. Ion exchange is a

a) Reversible process — b) Irreversible process
c) Vertical transport process — d) Horizontal transport process

13. ECEC is

a) > CEC — b) < CEC
c) = CEC — d) > AEC

14. AR_e^K is a measure of availability or intensity of

a) Labile pool of potassium in soil
b) Non- labile pool of potassium in soil
c) Fixed form of potassium in soil
d) None of the above

15. Geometric arrangement of the molecules is related with

a) Nitrogen fixation — b) Boron fixation
c) Phosphate fixation — d) Sulphate fixation in soils

Subjective

1. Define ion exchange phenomena in soil.
2. What are the dominant cations and anions in soil ?
3. What are the differences between CEC, AEC and ECEC ?

4. Outline the correlation between ARe^K and PBC^K of potassium in soil.
5. Differentiate between adsorption isotherms and adsorption envelope of ions in soil.
6. What is Schofield 's ratio law ?
7. How does water may act as ligand in soil ?
8. What are the conditions of ion-exchange in soil ?
9. What will be the ionic strength of 0.01 M $CaCl_2$ solution ?
10. What is the role of soil pH in ion exchange processes ?

Descriptive

1. What are the significance of ion exchange processes in soils ?
2. Outline in brief the factors affecting ion-exchange capacity in soil.
3. What are the different empirical equations related to ion exchange phenomena in soil ?
4. Briefly describe the Quantity -Intensity (Q/I) relation of potassium (K) in relation to K release in soil.
5. Outline the Donnan membrane equilibrium and its limitations in soil.
6. Describe the mechanism of ligand exchange in soil.
7. Outline the concept of adsorption isotherm and adsorption envelop at equilibrium in soil.
8. What is the importance of ion exchange phenomena in ionic environment in soil.?
9. What are the roles of base saturation in maintaining buffering capacity of soil.
10. What roles are played by chemical factors of soil in ion absorption by plants.

Suggested Readings

1. Sanyal, S. K., Poonia, S. R. and Barua, T. C. (2009). Soil colloids and ion exchange in soil. In: Fundamentals of Soil Science. (N. N. Goswami, R. K. Rattan, G. Dev, G. Narayanaswamy, D. K. Das, S. K. Sanyal, D. K. Pal and D. L. N. Rao eds.). Second edition. Indian Society of Soil Science, New Delhi.
2. Sposito, G. (1989). The Chemistry of Soils. Oxford Univ. Press
3. Beckett, P.H.T. (1964). Studies on soil potassium: 1. Confirmation of the ratio law: measurement of potassium potential. Journal of Soil Science, 15: 1-8.
4. Greenland, D.J. and Hayes, M.H.B. (1981). Chemistry of Soil Processes. John Wiley and Sons.

7

Soil Acidity, Alkalinity and Salinity

The soil acidity and alkalinity are major driving forces in regard to the mobility of plant nutrients from soil to plants. The extreme acidity or alkalinity cause damage to the crops and thus yield is affected. Correction of soil acidity or alkalinity depend upon the pH of the soil and its previous management options. Application of fertilizer in soil may be one of the factors affecting soil conditions. Exchangeable and soluble iron and aluminium ions and their hydroxy ion species may contribute hydrolysis to release H^+ ions to the soil solution. The soluble salts of carbonate and bicarbonates also develop salinity in soils.

7.0 Soil Reactions

Soil reaction is an indication of the acidity or alkalinity of soil and is measured in pH . The pH scale ranges from 0 to 14 with pH 7 as the neutral point. When the amount of active hydrogen ions in the soil solution increases the soil pH decreases and thus becoming more acidic in reaction. One of the outstanding physiological characteristics of the soil solution is its reaction. Soil reaction can influence the physico-chemical properties of soil which can regulate the growth and activity of plant and soil organisms associated with it.

7.1 Acidity

When the exchangeable bases from the surface layers of the soil are leached down, soil acidity is developed. The two adsorbed cations such as Hydrogen and Aluminium are largely responsible for soil acidity. In humid region, the soil acidity is dominant. When the hydrogen (H^+) ion becomes more than hydroxyl (OH^-) ion in the soil solution, the soil becomes acidic.

A soil may having the pH 4.5 will have low calcium and magnesium ions, high solubility of iron, manganese, aluminium etc., but low availability of nitrogen and phosphorus. The activity of microorganisms responsible for nitrification is adversely affected in acid soil. The acidic parent material (i.e. igneous rock) sometimes on weathering may develop acidity in soils.

Subsoil acidity

This kind of acidity is formed in highly weathered soil (ULTISOLS) restricting the crop yield even after correction of the sub-soil acidity by liming. The Al toxicity and unavailability of stored water in the rhizosphere could resist the root expansion. The leaching of surface applied liming materials could modify the sub soil acidity within 2-5 yeas. The Ca^{+2} from $CaSO_4$,$2H_2O$ leached down at a faster rate than from the $Ca(OH)_2$.

7.1.1 Active Acidity

This kind of acidity is due to the presence of H^+ and Al^{+3} ions in soil solution and constitute a very small part of total acidity.

7.1.2 Exchangeable Acidity

It is that part of total acidity which is replaceable by cation exchange with unbuffered salt solution ,such as KCl or simply the amount of exchangeable H^+ and Al^{+3} ions present on the soil surface.

7.1.3 Residual Acidity

It is associated with H^+ and Al^{+3} ions that are strongly bound in non-exchangeable form by clay colloids and organic matter and is not extractable by salt. Liming material is applied to neutralise the active, exchangeable and part of residual acidity for reclamation of acid soils.

7.2 Alkalinity

The soil that resembles pH more than 7.0 and having the adsorbed sodium to interfere with the growth of most crops is known as alkali soil. The adsorbed sodium ion can displace the calcium ions from clay colloid and makes the sodium- mixed clay particles. This sodium is converted into sodium hydroxide on hydrolysis as per the following reaction-

$$2Na^+ + CO_3^{-2} + 2H_2O \longrightarrow 2Na^+ + 2H^- + H_2CO_3$$

The OH^- ion thus formed increases the pH of the soil.

7.3 Neutrality

When the concentration of hydrogen and hydroxyl ion in soil solution is almost in equal quantities, the soils are neutral in character (pH 7.0).

7.4 Soil pH

The pH is the negative logarithm of the active hydrogen ion concentration of soil solution and is expressed in gram ion/lt or gram mol/lt. It is usually written as:

$$pH = \log \frac{1}{(H^+)} \text{ or } pH = -\log 10\ CH^+$$

pH scale is used to measure the acidity or alkalinity of a soil solution (or other solution). This scale runs from 0 pH to 14 pH, where the 7 units level is known as neutral, i.e., neither acidic nor alkaline. Pure water has a pH-7.0. All values below pH 7.0 denotes acidity and the values above pH 7.0 represents alkalinity. The degree of acidity increases as pH decreases below pH 7.0. If a soil having a pH 4.0 is ten times more acidic than showing pH 5.0 and is acidic in nature over that of a soil having a pH 9.0.

7.4.1 Factors Controlling Soil Reaction

i) **Nature of Soil Colloid:** When the site of soil colloids are dominated by adsorbed hydrogen (H^+) ions, the reaction of soil becomes acidic and when dominated by hydroxyl (OH^-) ion, the reaction of soil becomes alkaline in nature.

ii) **Nature of Ion:** The soil that contains more hydrogen ion than hydroxyl ions becomes acidic in reaction, even when there is presence of aluminium ions in the soil, they react with water to liberate hydrogen ions, which increases the soil acidity.

$$Al^{+++} + 3HOH \longrightarrow Al(OH)_3 + 3H^+$$

iii) **Percentage Base Saturation (PBS):** When there is a low percentage base saturation in soil, due to leaching of the basic elements from the soil (much below 80), the soil becomes acidic in reaction and on the other hand, when the PBS is above 80, the soils are likely to become alkaline in reaction.

iv) **Rainfall:** If the rainfall is high in a particular region , the soil becomes acidic in nature due to leaching of some nutrients such as calcium (Ca^{++}), magnesium (Mg^{++}) etc. from the soil solution. So, leaching encourages the development of soil acidity. On other hand, the soils that are developed in low rainfall areas, becomes alkaline in nature.

v) **Fertilizers:** The external addition of acid forming fertilizers (Ammonium sulphate, Urea, Ammonium nitrate etc.) in the soil in large quantities makes the soil acidic, while the basic fertilizers, such as Sodium nitrate, basic slag etc. makes the soil alkaline.

7.4.2 Sources of H^+ Ions in Soils

i) The exchangeable hydrogen ions are the main source of hydrogen ions in moderately acid soil (pH value between 5.0 and 6.5). The presence of aluminium in soil can react with water to liberate H^+ ions (as shown below) in soil solution on hydrolysis. At lower pH (4.0 to 4.7), due to higher concentration of H^+ in solution, the water molecules are not available to accept fresh proton generated on hydrolysis of Aluminium and as a result, different species of Al are produced (Figure 7.1).

$$Al^{+3} + HOH \rightleftharpoons Al(OH)^{+2} + H^{+}$$
$$Al(OH)^{+2} + HOH \rightleftharpoons Al(OH)_2^{+1} + H^{+}$$
$$Al(OH)_2^{+1} + HOH \rightleftharpoons Al(OH)_3 + H^{+}$$

$$Al^{+3} + 3HOH \rightleftharpoons Al(OH)_3 + 3H^{+}$$

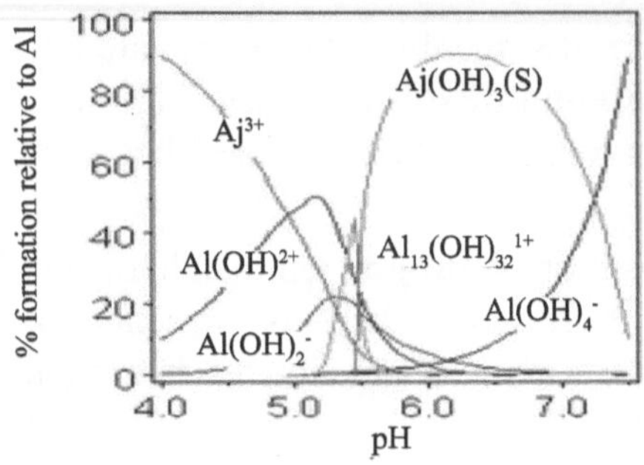

Figure 7.1. Effect of pH on Aluminium species

ii) Humus: The aluminum becomes less toxic by the presence of humus in soil. Soil pH will not change quickly in soils which are rich in organic matter. The soil organic matter sometimes may act as a buffer to reduce rapid acidification in soil by holding tightly the exchangeable H^+ in soils.

iii) Clay minerals: The H^+ ions present in the clay minerals can neutralise the negative charges in the tetrahedral /octahedral sheet or may also be released from the structural –OH groups present at the edges of the clay minerals and thus, pH-dependent charge (-ve charge) may be developed (Figure 7.1). The positive and negative charges may be developed on clay (Figure 7.2) on adjusting the pH of the soil solution.

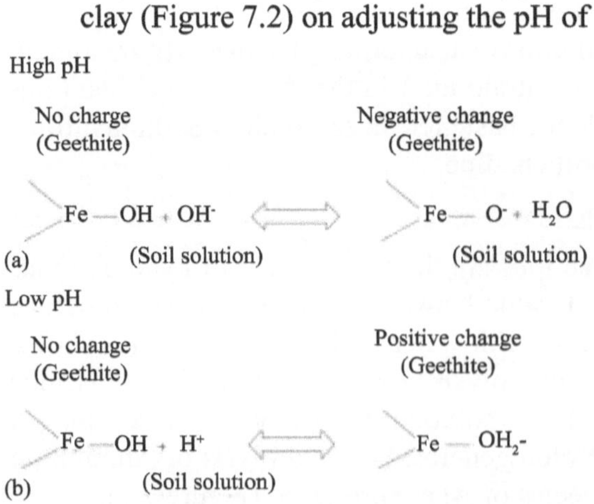

Figure 7.2. Development of charges at different pH of the clay

iv) Fertilizers: The nitrogenous fertilizers like ammonium sulphate, ammonium nitrate etc. when added in large quantity in soil can produce the nitrate that can combine with basic cations like calcium, magnesium and potassium and leach down from the topsoil into the subsoil. As these bases are removed or replaced by H+ ions, soils become more acidic.

v) Acid Rain: The kind of precipitation involving acidic components, such as sulphuric or nitric acid that fall on the ground from the atmosphere in wet or dry forms may be considered as acid rain or acid deposition. The acid rain results when sulfur dioxide (SO_2) and nitrogen oxides (NOX) are emitted into the atmosphere and are transported by wind and air currents. The SO_2 and NOX react with water, oxygen and other chemicals to form sulfuric and nitric acids and later fall on the ground as acidic precipitation. The major sources of SO_2 and NOX in the atmosphere are:

1. Burning of fossil fuels.
2. Vehicles and heavy equipment.
3. Oil refineries and other manufacturing industries.

Acid sulphate soil: In case of acid sulphate soils (pH < 3.5), the acidity is developed due to dissolved or free acidic substances, such as sulphuric acid, ferric and aluminium sulphate. The sulphuric acid is produced by oxidation of sulphur and sulphide. Acid sulphate soils occur in some of the low lying areas of Kerala, south east coast of Tamil Nadu, Sundarbans of WestBengal and mangrove areas of Andaman Islands.The pH of the soil sometimes becomes close to pH 4.5 or even less (pH < 4.0).

Potential Acid Sulphate Soil

In this soil both sulphur and iron are found in reduced state and the soil become acidic on draining water. A strong smell of H_2S is found from the soil when disturbed the presence of pyrite in the soil and sediment could act as a potential acid sulphate soil when the potential acidity exceeds the soil's acid neutralizing capacity ($CaCO_3$). The acid sulphate soil will have a pH 4.0 under aerobic condition and is also associated with yellow mottles or coating of Jarosite [$KFe_3(OH)_6(SO_4)_2$] which are formed on reaction of H_2SO_4 with iron bearing minerals.

7.5 Genesis of Acid Soils

There are different factors which are responsible for the formation of acid soils.

a) Climate: In high rainfall areas , excess water moves downward with soluble salts (basic cations) present in the soil surface and the surface soil becomes acidic in reaction. The compounds of iron and aluminium are resistant to leaching and impart acidity to surface soil (laterite soil). It has been observed that, with an annual rainfall > 1300mm can have soil pH around 5.0.

b) Vegetation: The regions endowed with ample forest cover or with vegetation can have acid soil. The leaf litter of the trees undergoes decomposition and impart acidity to the soil.

c) Parent materials: It has been reported that, the soils developed from the acidic parent materials (Igneous Rock), can lead to acid soil. The alkaline salts can be percolated by the water passing through the perforation of the rocks which are quite soft.

d) Topography: The acid soils can be developed in sloppy areas and not in the river basin. The regions (plain) with good drainage facility also can develop acid soil in future.

e) Human interferences: The indiscriminate use of irrigation water with improper drainage facility and sometimes the continuous use of acid forming fertilizer (ammonium sulphate, ammonium chloride etc.) can lead to development of soil acidity.

7.6 Active and Potential acidity

Active acidity is the amount of hydrogen ions present in the soil solution at any given time. Reserve acidity is the amount of acid forming ions being held by the soil colloids. Sometimes, the reserved acidity is called the potential acidity.

Total acidity = Active acidity + Reserved /Potential acidity

The exchangeable acidity is sometimes referred to as "reserve acidity". The two types of acidity therefore exist in an equilibrium relationship. The amount of active acidity determines the pH of the soil solution at any given time. The quantity of H^+ ions contributing to active acidity is, however, very small compared to exchange acidity.

Determination of soil pH: The soil pH in general is determined by maintaining a soil : water ratio of 1:2.5 and the suspension thus made is measured in a pH meter (Figure 7.3) after stirring for 15-20 minutes. The pH of the soil suspension will vary with the changing soil: water ratio.

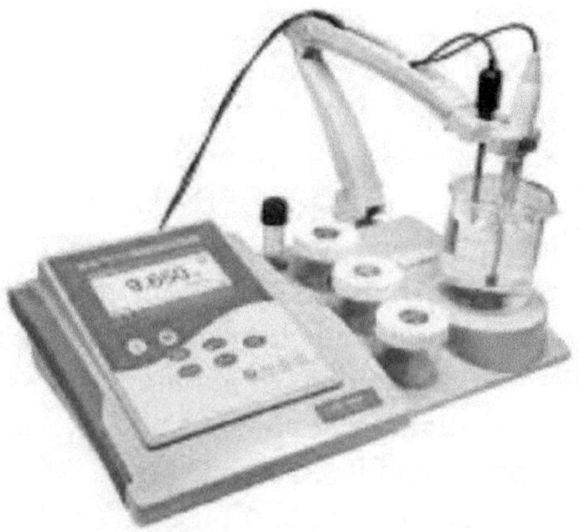

Figure 7.3. Digital pH-meter

Lime requirement: The term "lime requirement" may be defined as the amount of lime ($CaCO_3$) or its equivalent that must be applied to soil to increase the pH to 7.0 or some prescribed value. Considering the pure $CaCO_3$ as standard with 100% neutralising power, the neutralising power of the other liming materials [CaO, $Ca(OH)_2$, Ca,Mg $(CO_3)_2$, $CaSiO_3$] can be determined by the following formula :

Equivalent weight of $CaCO_3$ x Neutralising value(NV) of $CaCO_3$ =Equivalent weight of the limimg material x its Neutralising Value (NV)

Or, NV or CCE (calcium carbonate equivalent) of liming material = (Equivalent weight of $CaCO_3$ x NV of $CaCO_3$) /Equivalent weight of liming material

Lime potential: It was suggested by Schofield and Taylor (1955) that, instead of using ion activity, it is better to use ion activity ratio for determination of soil acidity. When an exchange complex is saturated with H^+ and Ca^{+2} ions, then according to the ratio law at equilibrium (Schofield) :

$[H^+]/ [\sqrt{Ca^{+2}}] = \text{Constant}$ (7.1)

The equation,

pH-1/2 pCa = Constant (7.2)

is known as lime potential (LP) which (Eq 7.2) is unaffected by the ratio of soil: water and salt concertation and reliable character of any soil *vis- a- vis* soil pH.

$$LP = [H^+]/[Ca^{+2}] = \text{Constant}$$

$$\text{or, pH-1/2 pCa} = \text{Constant}$$

$$\text{or, pH} = 1/2\text{pCa} + \text{Constant}$$

7.7 Electrochemistry of clays

It requires to convert the heterogeneity of soil-clay system into a homoionic system to understand the nature and properties of soil-clays. The base unsaturation of clay by adding 0.1(N) HCl and subsequent removal of free HCl by water results the H-clay systems. The exchange capacity of H-clay may be determined by titrating with an alkali through electrometric measurements. The weak acid characteristics may be produced both in potentiometric and conductometric measurements, although ,the strong acid characteristic curves may be shown in presence of neutral salts and the amount of acidity will also be higher in presence of salts. The two phase systems are found in acid clay; the insoluble acidic material consisting of one phase and interparticle liquid being the other. The H^+ or H_3O^+ ions become the part of the solid phase. Besides, the CEC or total reacting acid of H-clay or clay acid varies according to the pH at which it is measured. There may be quantitative differences of titration curves of the same clay acid with different electrolytes (bases). On addition of neutral salts to acid clay, H^+ ions are liberated by cation exchange reactions along with Al^{+3} ions which increases with the concentration of the added salts and thus, the acidity of a clay acid suspension is equivalent to the total quantity of exchangeable H^+ and Al^{+3} ions (Figure 7.4). In presence of the addition of neutral salts, the characteristics of the titration curves change from weak acid to strong acid. Hence, two types of associated H^+ ions are observed in clay acid (Biswas and Mukherjee, 1994); one titratable with the base and another titratable in presence of added salts (Neutral). The di or tri-basic characteristics may be observed when more than one inflexion points are found in the titration curves.

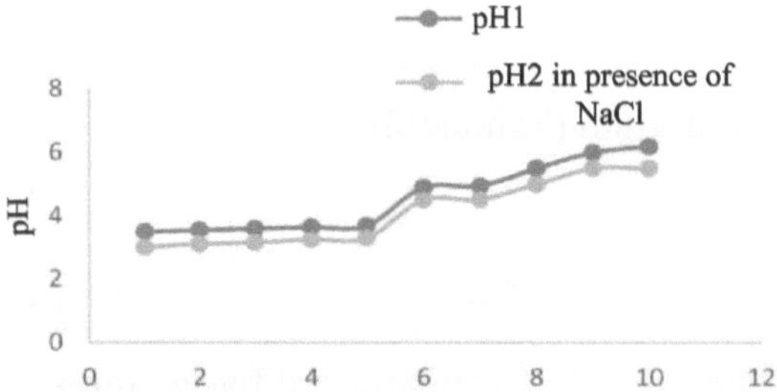

Figure 7.4. Effect of salt on potentiometric titration of acid clay

7.8 Management of Acid Soils

i) Use of liming materials: Incorporation of lime neutralizes soil acidity, increases microbial activity, nutrient availability and thus improves the physical conditions of soil. Among the naturally occurring lime sources, calcite and dolomite or paper mill sludge are widely used for correction of soil acidity.

ii) Nutrient management: Application of organic manures in combination with inorganic fertilizers and lime proves to be an excellent package for improving productivity and health of acid soils. The problem of P - deficiency in the acid soil has to be tackled largely through the application of adequate amount of suitable type of P - fertilizer and by adoption of proper management practices. Rock phosphate can be used in legumes in *Rabi* season and the residual effect may be utilized in the following *kharif* upland crops. Application of indigenous rock phosphate as a source of P in long duration crops (eg. horticultural crops) is effective. The Fe- hazard in soils can be minimised by improving drainage facility around the fields by constructing check embankments across the slope with provision of diversion weirs and liming materials may be used to check iron toxicity in soil.

iii) Crop Scheduling: Under the rainfed conditions, highly responsive crops like cotton, soybean, pigeon pea, groundnut, frenchbean etc. may be considered for growing in the first year of liming which could be followed by medium responsive crops like maize and wheat in the subsequent seasons. Cultivation of low responsive crops (millets, rice, barley, linseed etc.) may be grown in the last season/ year prior to commencement of the subsequent liming schedule.

iv) Diversification of crops: Crop diversification in respect to tolerant crops like (tea, groundnut, maize etc.) may be considered.

v) Selection of varieties : In acid soil, rice is traditionally grown irrespective of the type of land (Upland, Medium land and low land) as rice has certain degree of tolerance to soil acidity. On the other hand, cultivation of highly sensitive (towards soil acidity) field crops (soybean, french bean, pigeon pea etc.) and or horticultural crops (mango,banana.peach etc.) should be framed accurately to liming and/or other ameliorative measures.

vi) Use of organic manures: Incorporation of decomposed organic matter in acid soils can prevent sudden fluctuation of soil pH by improving the buffering capacity of soils. Moreover, it increases the availability of P and reduces the toxicity of Fe and Al in acid soils. Fresh mulches (mostly weed biomass) also reduces the adverse effect of soil acidity substantially.

7.9 Soil reaction and Plant Growth

Soil reaction is associated with the pH of the soil which actually refers to the relative acidity or alkalinity in the soil of a particular region. It is directly proportional to the amount of nutrients that will be absorbed by the plants.

The optimum availability of nutrients for most crops is 6.5 pH to 7.0 pH. If the pH is below 6.0 some nutrients, such as nitrogen, phosphorus, and potassium, are less available, while when the pH exceeds 7.5, iron, manganese are less available. Plants can grow in a pH of slightly acidic to neutral soil pH (6.8 -7.2). The pH level of soil is influenced by the environmental factors of vegetation types, precipitation and temperature. The area under the high rainfall and forest cover, has a moderately acidic soil. On the contrary, drought-prone regions have an alkaline pH level, while moderate rainfall have neutral pH value. However, there may be changes in pH of the soils by the adoption of management practices. Plants like blueberries, potatoes and rhododendrons etc. can be grown in moderately acidic to neutral pH. A plant growing in extremely acidic soils may exhibit symptoms of aluminium (Al), hydrogen (H) or manganese (Mn) toxicity. It may suffer from potential nutrient deficiencies of calcium and magnesium. It has been observed that nutrients required for the plants can be dissolve easily at the pH of the soil solution ranging from 6.0 to 7.5.

7.10 Alkaline Soil

The alkalinity of a soil is developed due to rise in pH more than 7.0 by the presence of Na^+, Ca^{+2}, Mg^{+2} etc. Total soil alkalinity of a soil increases with:

i) Weathering of silicate, aluminosilicate and carbonate minerals containing Na^+, Ca^{2+}, Mg^{2+} and K^+

ii) Addition of carbonate minerals to soils eroded elsewhere by wind or water, or by mixing of the soil with less weathered material (such as the addition of limestone to acid soils).

iii) Addition of water, containing excessive amount of dissolved bicarbonates (as occurs while irrigating with high-bicarbonate containing waters).

iv) The development of alkalinity in a soil (due to the deposition of carbonates and bicarbonates of Na^+, K^+, Ca^{++} and Mg^{++}) occurs when there is insufficient water flowing through the soils to leach soluble salts. In arid situation, if the drainage system is poor or the water entering into the soil get evaporated , there is deposition of alkaline salts on soil.

v) It has been observed that, increasing the amount of sodium in an alkaline soil tends to induce dissolution of calcium carbonate, which rather increases the soil-pH. The calcareous soils may vary in pH from 7.0 to 9.5, depending on the degree of relative dominance of Ca^{+2} and Na^+ as soluble cations.

7.10.1 Types of Alkali Soil

There are two types of alkali soils based on the dominance of Na^+ ions.

i) Black alkali soils—In these soils, Na_2CO_3 is found in excess. Bertholet suggested that Na_2CO_3 was formed in black alkali soil by interaction of NaCl and $CaCO_3$.

ii) White alkali soils—In this group, NaCl is present in excess. Russians call such soils as solonchack.

7.10.2 Development of Alkaline Soil

Followings are some of the reasons for development of alkaline soil.

1. The natural cause of alkaline soil is the presence of soil minerals producing sodium carbonate (Na_2CO_3) and sodium bicarbonate ($NaHCO_3$) upon weathering.
2. Using coal or lignite rich limestone in the coal fired boiler or power plants could produce ash containing calcium oxide which when mix with soil can increase the soil alkalinity.
3. The Sodium salts are used in industrial and domestic applications in huge quantities, may enter into the river / groundwater during their production process or consumption and enhance water-sodicity. This water rich with soluble sodium when used for irrigation could increase the sodicity in soil and the soil become alkaline.
4. The application of softened water in irrigation containing high proportion of sodium bicarbonates and less calcium and magnesium cause the alkalinity in soil.

Besides, the soil alkalinity is associated with the presence of sodium carbonate (Na_2CO_3) or sodium bicarbonate ($NaHCO_3$) in the soil, either as a result of natural weathering of the soil particles or brought in by irrigation and/or floodwater. Sodium carbonate in water can produce carbon dioxide (CO_2), which may escape as a gas or absorbed by Algae and sodium hydroxide (Na^+OH^-), which is alkaline (or rather basic) and generates high pH values (pH>8.5).

$$Na_2CO_3 + H_2O = 2NaOH + CO_2\uparrow$$

$$NaHCO_3 = NaOH + CO_2\uparrow$$

7.10.3 Reclamation of alkaline soil

The pH of an alkaline soil can be reduced by adding acidifying agents or organic (peat or peat moss) materials. Sometimes, to reduce the excessive alkalinity in soil, elemental sulfur (90–99% S) are used @ 300–500 kg/ha – it slowly oxidizes in soil to form sulfuric acid. Acid producing fertilizers, such as ammonium sulfate, ammonium nitrate and urea can help to reduce the pH of a soil when ammonium is oxidised to form nitric acid. However, in high-pH soils with a high calcium carbonate content (more than 2%), it is advised to add phosphorus, iron, manganese, copper and/or zinc, because deficiencies of these nutrients are the most common reasons for poor plant growth in calcareous soils. Again, gypsum (calcium sulphate) can also be applied as a source of Ca^{++} ions to replace the sodium at the exchange complex. Gypsum also reacts with sodium carbonate to convert into sodium sulphate which does not contribute to high pH. There must be enough natural drainage to the underground, or else an artificial subsurface drainage system must be present to permit leaching of the excess sodium by percolation of rain and/or irrigation water through the soil profile. Calcium Chloride is also used to reclaim alkali soils by converting Na_2CO_3 into NaCl and thus precipitating $CaCO_3$, while NaCl is drained off by leaching water.

$$CaCl_2 + Na_2CO_3 = 2NaCl + CaCO_3$$

Alkali soils are also reclaimed by the cationic exchange (replacement of sodium from soil colloids by calcium ions) reaction; Application of calcium sulphate (gypsum) in the soil reduces alkalinity to a great extent and makes the soil fertile.

Reactions

$$2\text{Na-clay} + CaSO_4 = Na_2SO_4 + \text{Calcium-clay}$$

$$Na_2CO_3 + CaSO_4 = CaCO_3 + Na_2SO_4 \downarrow$$

The Na_2SO_4 is leached down where there is good drainage facilities.

The salt percentage in an alkali soil can also be reduced by the use of acid forming substances such as sulphur, ferrous sulphate and limestone. Sulphur, when applied to the soil, oxidises and forms sulphuric acid which converts carbonates of sodium and potassium to Na_2SO_4 and K_2SO_4 respectively that may be removed from the surface soil by the drainage water. The amount of gypsum and sulphur required to ameliorate the alkali soils depends on the degree of alkalinity, drainage and buffering capacity of soils.

Reactions in an alkaline soil:

1. With Sulphur

i) $2S + 3O_2 = 2SO_3$

ii) $SO_3 + H_2O = H_2SO_4$

In the calcareous soil,

iii) $H_2SO_4 + CaCO_3 = CaSO_4 + CO_2 + H_2O$

iv) $2Na\text{-}Clay + CaSO_4 = Ca\text{-}Clay + Na_2SO_4$

In non-calcareous soil,

v) $2Na\text{-}Clay + H_2SO_4 = 2H\text{-}Clay + Na_2SO_4$

2. With lime-sulphur

$CaS_5 + 8O_2 + 4H_2O = CaSO_4 + 4H_2SO_4$

In non-calcareous soil,

$10Na\text{-}Clay + 4H_2SO_4 + CaSO_4 = 8H\text{-}Clay + Ca\text{-}Clay + 5\ Na_2SO_4$

3. With Ferrous sulphate

i) $FeSO_4 + H_2O = H_2SO_4 + FeO$

In calcareous soil,

ii) $H_2SO_4 + CaCO_3 = CaSO_4 + CO_2 + H_2O$

iii) $2Na\text{-}Clay + CaSO_4 = Ca\text{-}Clay + Na_2SO_4$

In non-calcareous soil,

$2Na\text{-}Clay + H_2SO_4 = 2H\text{-}Clay + Na_2SO_4$

4. With limestone on non-calcareous soils

i) $Na\text{-}Clay + H_2O = H\text{-}Clay + NaOH$

ii) $2H\text{-}Clay + 2NaOH + CaCO_3 = Ca\text{-}Clay + Na_2CO_3 + 2H_2O$

5. With H-Clay

$2H\text{-}Clay + CaCO_3 = Ca\text{-}Clay + CO_2 + H_2O$

6. Reaction with Pyrite (FeS_2)

$FeS_2 + 2Fe^{+3} = 3Fe^{+2} + 2S$

Sulphur thus formed could be the substrate for thioxidants which converts it into H_2SO_4.

$S + 3(O) + H_2O = H_2SO_4$

$FeS_2 + H_2O + 7O = FeSO_4 + H_2SO_4$

$2FeSO_4 + O + H_2SO_4 = Fe_2(SO_4)_3 + H_2O$

$FeSO_4$ formed in the above reaction may be converted into H_2SO_4 by hydrolysis.

$FeSO_4 + H_2O \rightarrow H_2SO_4 + FeO$

Thus, the pyrite is oxidized in soils to ferrous sulphate and sulphuric acid:

$2FeS_2 + 2H_2O + 7O_2 = 2FeSO_4 + 2H_2SO_4$

Both sulphuric acid and ferrous sulphate help in reclamation of calcareous as well as non- calcareous soils affected with soluble salts by lowering the pH and solubilising free calcium from calcium carbonate present.

7. In salt-affected calcareous soils

$CaCO_3 + H_2SO_4 = CaSO_4 + CO_2 + H_2O$

$2Na\text{-}Clay + CaSO_4 = Ca\text{-}Clay + Na_2SO_4$

$FeSO_4 + H_2O \rightarrow FeO + H_2SO_4$

$2NaHCO_3 + H_2SO_4 \rightarrow Na_2SO_4 + 2H_2O + 2CO_2$

$Na_2CO_3 + H_2SO_4 \rightarrow Na_2SO_4 + H_2O + CO_2$

8. When the soil is non-calcareous

i) $2Na\text{-}Clay + H_2SO_4 = 2H\text{-}Clay + Na_2SO_4$

ii) $FeSO_4 + H_2O = FeO + H_2SO_4$

The H_2SO_4 thus formed in reaction will act in similar way as before. Sometimes, the alkalinity and salinity of the soil are reduced by the use of molasses and press-mud.

7.11 Alkaline Soil and Plant Growth

The alkaline soil can affect the plant growth and nutrition due to the presence of excess soluble salts in rhizosphere.

i) The excessive concentration of salts in the soil solution decreases absorption of nutrients by plants and causes plasmolysis of cell cytoplasm which may cause the stunted growth of plants.

ii) The excess absorption of sodium or chloride salt of alkaline elements become toxic to the plants and may be the cause for the death of plants.

iii) Sometimes, the presence of excess salts in the soil retards the germination of seeds and growth of seedlings and plants will die before fruit settings.

iv) The size of leaves is reduced in alkali soil, plant roots remain superficial, bark of stem turns brown or black, green tissues are less developed and ultimately the growth of the plants is restricted.

7.11.1 Alkali -tolerance Plants

Some plants are resistant to alkali salts. Crops like Barley, wheat, oats, sorghum, sugar-beet, berseem are best suited to grow in alkaline soil. Cotton and grapes are also alkali tolerant. High salt tolerant crops are Dhaincha, paddy, sugarcane in *Kharif* and oat, berseem, lucerne, sanji (*Trigonella* spp), and barley in *Rabi*. The medium salt tolerant crops which could be tried during second stage of reclamation are castor, cotton, jowar, bajra and maize for *Kharif* and mustard and wheat for *Rabi* season. The low salt tolerant crops are sesamum, moong, urd, arhar and sunhemp in *Kharif* and gram peas, linseed during *Rabi*. Besides, the trees like babul, dhak, jhand, khair, chokra, neem, Lasora, Sisham, siris, bahera and reonjha can be planted on saline- alkali soils.

7.11.2 The alkali tolerance of plants depends upon

i) The physiology of the plants.

ii) The shallow roots are more affected by alkalinity than the deeper roots.

iii) The salts present in the alkaline soil can reduce acidity and improve the physical conditions of the soils.

iv) Calcium salts provide calcium to the plants.

v) In alkaline soil, the toxicity of aluminium and Mn may be reduced.

7.12 Salt Affected Soils

The salt-affected (salinity or alkalinity) soils are developed for the following reasons:

i) Arid climate: The salt accumulation is commonly found under arid climatic situation. In India, about 25 million hectare land is salt affected.

ii) Elevated subsoil water table: When the groundwater table is within the capillary range, the water withsoluble salts could rise to the soil surface. The salts are deposited on the surface for evaporation of water.

iii) Poor drainage: Due to the poor drainage systems, accumulation of water in soil creates water logging which leads to salt accumulation.

iv) Quality of irrigation water: Irrigation water containing more than permissible quantities of soluble salts (sodium carbonate and bicarbonates) make the soil saline.

v) Sea water inundation: In coastal area, periodical inundation of land by sea water during hightides make the soil salty.

vi) Nature of parent materia: The saline nature of parent rock leads to salt accumulation.

7.12.1 Characteristics of the salt-affected soil

Parameters	Saline soil	Non-saline Alkali soil	Saline-alkali soil	Degraded alkali soil
Salt content	Excess soluble salts of sodium exchangeable calcium.	Absence of soluble salts, presence of excess exchangeable sodium in soil matrix.	Both saline and non-saline alkali soils in nature, contains sodium and soluble salts	Presence of H^+ ions in the upper layer and sodium (Na^+) in the lower layer
Colour	White on soil surface	Black	—	Black in the lower layer
Sodium adsorption ratio (SAR)	Less than 13	More than 13	More than 13	Less than 13 in the surface and more than 13 in the lower layer
Exchangeable sodium percentage (ESP)	Less than 15 of total CEC	More than 15 of the total CEC	More than 15 of the total CEC	More than 15 of the total CEC
Soil pH	Less than 8.5	8.5-10.0	More than 8.5	pH nearing 6.0 in the surface and 8.5 in the lower layer
Physical condition of soil	Soils are mostly flocculated and permeable to water and air.	Deflocculated condition, poor permeability to water and air.	Salts of sodium content will determine the flocculation and deflocculation of soil	Low infiltration and permeability
Organic matter content	Slightly less than normal soils	Very poor due to the presence of sodium carbonate	Variable	Low
Total soluble salts (TSS)	More than 0.1%	Less than 0.1%	More than 0.1%	Less than 0.1%
Electrical conductivity of the saturation extract (ECs)	>4 dSm^{-1}	<4 dSm^{-1}	>4 dSm^{-1}	<4 dSm^{-1}

7.12.2 Reclamation of the Salt-Affected Soils

1. Chemical amendments

Amendments	Soil conditions
Gypsum	Saline and alkali soil having pH up to 9.0
Sulphur, iron sulphate, iron pyrite	Alkaline and saline-alkali soils having pH ranging from 8.0-9.0
Limestone	Saline soils having pH < 8.0

2. Subsurface drainage, leaching or flushing

The following measures may be taken for reclamation :

a) Establishing the internal drainage system (e.g. tile drains, mole drain, ditching etc.).

b) Replacing the excess exchangeable Na^+ ions.

c) Leaching out most of the soluble salts from the root zone by supplying salt free irrigationwater.

d) Addition of organic mulch in the soil.

For the reclamation of sodic or saline-sodic soils, the sodium ions must be replaced from the exchangeable sites, here, Gypsum ($CaSO_4$, $2H_2O$) may be applied.

In salt -affected calcareous soils

$$CaCO_3 + H_2SO_4 = CaSO_4 + CO_2 + H_2O$$

$$2Na\text{-}Clay + CaSO_4 = Ca\text{-}Clay + Na_2SO_4$$

$$FeSO_4 + H_2O \rightarrow FeO + H_2SO_4$$

$$2NaHCO_3 + H_2SO_4 \rightarrow Na_2SO_4 + 2H_2O + 2CO_2$$

$$Na_2CO_3 + H_2SO_4 \rightarrow Na_2SO_4 + H_2O + CO_2$$

Gypsum requirement (GR): The amount of gypsum required to be added to a sodic soil to lower the ESP to the desired value is known as Gypsum requirement and is expressed as Ca^{+2}/100g soil.

GR = [ESP (Initial)-ESP (Final)] $\times$ CEC/100 and expressed as me of Ca^{+2}/100g soil

Where, ESP is the exchangeable sodium percentage obtained from soil analysis before reclamation (ESP initial) and after reclamation (ESP final). CEC = Cation Exchange Capacity in cmol (p+) kg^{-1}.

7.12.3 Some Appraisal of Saline and Sodic Soils

1. Osmotic pressure (O.P.): The O.P. is calculated from the freezing point depression in soil solution

O.P. = 12.06 ΔT – 0.021 ΔT^2

Where, ΔT = depression of freezing point

The osmotic pressure is regulated by the water uptake and the growth of the plants in saline soilAgain, O.P. = 0.36 × EC

where, EC= Electrical conductivity (dSm^{-1})

2. Water soluble boron and saline soils

Boron concentration (ppm)	Crop growth
<0.7	Safe
0.7-1.5	Marginal
>1.5	Unsafe

3. Electrical Conductivity and Crop Growth

ECs (dSm^{-1})	Crop growth
< 2	Effect is negligible
2-4	Yield of very sensitive crops restricted
4-8	Yield of some crops restricted
8-16	Tolerant crops can survive effectively
>16	Only a few crops can survive effectively

4. Sodium adsorption ratio (SAR): The equilibrium condition between the soluble salts and exchangeable cations may be ascertained:

$$SAR = Na+/\sqrt{}\,(Ca^{+2} + Mg^{+2})/2$$

Na^{+1}, Ca^{+2} and Mg^{+2} (me l^{-1}) are in the saturation extract

Again, ESP =100 (-0.0126 + 0.01475 SAR)/1+ (-0.0126 + 0.01475 SAR)

ESP = Exchangeable Sodium Percentage of the saturation extract

Again,

Y = 0.0673 + 0.035X

Where, Y = ESP; X = SAR

7.12.4 Management Options

The saline soil can be easily improved by leaching of salts using good quality water and also by good drainage facilities.

i) Application of gypsum would improve the permeability of soil by making good soil aggregates

ii) Growing of resistant crops like ragi cotton, barley and rice

iii) Growing green manure crops like sunnhemp, daincha and kolinji

iv) The encrustation of salts could be alleviated by incorporating organic matter and by adding montmorillonite containing silt.

Questions

Objective

1. What is the ionic form of phosphorus when soil pH is more than 9.0 ?
   a) $H_2PO_4^-$ b) HPO_4^{--}
   c) PO_4^{---} d) H_3PO_4
2. Which liming material is a source of magnesium ?
   a) Calcitic limestone b) Dolomitic limestone
   c) Calcium hydroxide d) Calcium oxide
3. If pH reading is zero for any suspension, what is its H^+ ions concentration in mol/L
   a) 1 b) 2
   c) 3 d) 4
4. Neutralizing value (%) of dolomite is
   a) 179 b) 136
   c) 86 d) 108.7
5. Which of the following ions are common in acidic soils?
   a) Ca^{2+} b) Al^{3+}
   c) Mg^{2+} d) OH^-
6. Phosphate fixation is prominent in
   a) Acid soil b) Alkaline soil
   c) Saline soil d) Salaine-alkaline soil
7. The structural formula of monocalcium phosphate is
   a) $CaHPO_4.2H_2O$ b) $CaHPO_4$
   c) $Ca_3(PO_4)_2$ d) $Ca(H_2PO_4)_2.2H_2O$
8. Podzolisation is
   a) Negative of calcification
   b) Removal of sesquioxide from A horizon
   c) Both a & b
   d) None of the above

9. Liming material passing through a 60 mesh sieve has efficiency (%)
   a) 60 b) 80
   c) 90 d) 100
10. CCE (%) of Basic slag is
   a) 100 b) 179
   c) 86 d) 136
11. The optimum pH for nitrification process is
   a) <5.0 b) 5.0-6.0
   c) 6.5-7.5 d) 7.5-8.5
12. Solonchak is formed due to
   a) Salinization b) De-salinization
   c) Solonization d) Laterization
13. For reclaiming acid soils, we use:
   a) Gypsum b) Lime
   c) Profile inversion d) Sub-soiling
14. Saline –alkaline soil has EC value
   a) > 4 mmhos cm^{-1} b) < 4 mmhos cm^{-1}
   c) 2 mmhos cm^{-1} d) 1.5 mmhos cm^{-1}
15. Black alkali soil has
   a) High Ca with Na
   b) More organic matter
   c) Dominance of carbonate anions
   d) Less Na than other alkali cations

Subjective

1. What do you mean by soil acidity and alkalinity.
2. What is total acidity in soil ?
3. What are the chemicals used for correcting soil acidity and alkalinity?
4. Name the acid tolerant and alkali tolerant plants.
5. What is sub-soil acidity ?
6. How does Al^{+3} help formation of Podsol soil ?
7. What are the EC,ESP and pH of a saline soil ?
8. How soil acidity is measured in the laboratory ?

9. What are the roles of Al^{+3} /Fe^{+3} in producing soil acidity ?
10. What are the characteristics of sodic soil ?

Descriptive

1. What are the causes of developing soil acidity ?
2. What are the factors affecting soil reaction in acid soil ?
3. How H-clay or acid clay is prepared in the laboratory ?
4. Outline the characteristics of potentiometric titration curves of an acid clay.
5. What are the different fractions of total acidity in soil ?
6. How acid soil is reclaimed ?
7. What are the characteristics of salt affected soil ?
8. How alkali soil is reclaimed ?
9. What are the influences of soil acidity and alkalinity on plant growth and nutrition ?
10. How is degraded alkali soil produced ? What are the process of rejuvenation of degraded alkali soil ?

Suggested Readings

Brady, N.C. and Weil, R.R. (2007). The nature and properties of soils. 14th Edition, Prentice Hall, USA.

Mohsin, M.A., Sarkar, A.K. and Mathur, B.S. (1995). Acid soil management. Kalyani Publishers.

Pal, S.K. (2016). A textbook of soil science. Oxford and IBH Publishing Co. Pvt. Ltd., New Delhi.

Schofield, R.K. and Taylor, A.W. (1955). The measurement of soil pH. Soil Sci. Soc. Am. Proc. 19: 164-167.

8

Fixations of Nutrients

Nutrient adsorption in soil is one of the important phenomena in terms of nutrient availability to plants. Due to the nutrient fixation of the added nutrients supplied through fertilizers or manures may result the increase in cost of production due to the non-availability of the added nutrients at the optimum level to the plants. Hence, management options are required to release the sorbed nutrients from soil for better availability to the plants. The ion-fixation is driven by physico-chemical and biological processes and varies under agro-ecological situations. The chemistry of ion fixation in the form of specific or non-specific adsorption on soil colloids will help understanding the bonding and adsorption reactions of ions in the soil.

8.0 What is Fixation?

Fixation of plant nutrients in soil may be defined as the process, where readily soluble plant nutrients are transformed into less soluble forms by physical or chemical reaction with organic or inorganic compounds and thus restricting the mobility of ions in soil that leads to the decrease in availability to the plants. There are two kinds of fixation in general, viz; cation (N, K, Fe, Mn, Cu, Zn, Ca,Mg etc.) and anion (P, B, Mo etc.) in soils. The fixation of ions in soil is governed by important physical and chemical parameters of soil based on which the available pools of nutrients for plants may be determined.

Cation fixation: Cation fixation in soils is possible when water-soluble or exchangeable cations are converted to unavailable forms to the plants and can not be readily extracted with the commonly used extractants. The cations like NH_4^+ and K^+ are basically involved in such reaction processes.

8.1 Mechanism of NH_4^+/K^+ Fixation

The fixation of NH_4^+ occurs in the crystal lattices under different clay minerals (i.e., Montmorillonite, Illite and Vermiculite). Sometimes amorphous oxy and hydroxide (Fe and Al) and non-crystaline allophone minerals also adsorb NH_4^+ which may subsequently be transformed into other minerals. The ammonium containing fertilizers when applied into the soil could react with soil organic matter to form undecomposable compounds and is said to be fixed by the organic substances, possibly by the quinones or phenolic –OH groups present in the organic matter.

NH_4^+		NH_4^+		NH_4^+
	⟷		⟷	
K^+		K^+		K^+
[Solution]		[Exchangeable]		[Non-exchangeable or fixed]

The fixation processes of NH_4^+ are similar to that of K^+ fixation as for the close ionic sizes and degree of polarization of the later (K^+).

The "Lattice hole theory" was proposed by Page and Baver (1940) on K-fixation considering the following criteria:

1. Expanding - contracting nature of 2:1 type clay minerals
2. The geometric arrangement of oxygen ions at the interlayer position of the expanding type (2:1) of clay minerals
3. The size of the unhydrated ions.

The sheet of oxygen ions arranged hexagonally at the outer surface of the 2:1 minerals forms an opening of 2.8 A° in diameter at the interlayer position. The surroundings water hull of the ions are lost during dehydration and the layers are contracted having the ionic diameter (K^+, 2.66A° ; NH_4^+, 2.86 A°) close to the ionic diameter of 2.80 A°, which enables the ions(K^+, NH_4^+) to fit snugly into the cavities and also to make the layers close together against further rehydration and re-expansion. The larger cations can not enter into the cavity and the smaller being too small to contract and bind the two layers together.

8.2 Factors Affecting NH_4^+/K^+ Fixation

i) Soil colloids: Potassium ion is fixed normally in 2:1 type of clay minerals where the order of fixation is Vermiculite>Illite>Smectite, depending upon the sources of negative charge and charge density of clay. Whatever the excess negative charge is generated at the outer surface of the clay minerals are balanced by the K^+ ions between the layers. It has been observed that, lowering of the c-axis spacing will attract more K^+ in non-exchangeable form in the order as: Illite (10A°) > Vermicullite (14A°) > Montmorillonite (18A°).

ii) Alternate Wetting and Drying/ Freezing and Thawing: The release or fixation of K^+ in soil depends on the stock of K^+ in the soil. When the K-content in soil is low, alternate wetting and drying/ freezing and thawing favour the partial release of non-exchangeable K to exchangeable or solution form and when the available pool of potassium in soil is high, the process of alternate wetting and drying/ freezing and thawing causes fixation of some available K to non-exchangeable form.

iii) Soil reaction and pH: At low pH, chances of K-fixation is reduced due to the presence of hydroxy-aluminium $[Al(H_2O)_6]^{3+}$ and hydronium (H_3O^+) ions having larger ionic diameter compared to the hexagonal cavity. These ionic species are replaced by increasing pH with liming materials from the interlayer position and makes the non-exchangeable sites free for K- fixation. The liming favours the fixation of NH_4^+ also.

iv) CEC of soil: With the increase in CEC and percentage base saturation in soil (PBS), the fixation of both K^+ and NH_4^+ will increase.

v) Moisture: At low concentration of NH_4^+ in soil, the fixation of NH_4^+ is usually increased by drying, while even after the added NH_4^+, the fixation may decrease with an increase of water : soil ratio.

vi) Organic matter: The organic matter helps in fixing NH_4^+ in soil. The fixation process by organic matter for NH_4^+ is more in alkaline soil than acidic one.

8.3 Losses of Nitrogen from Soil: Nitrogen is lost from soil by

a) Crop removal: In true sense the nitrogen is assimilated by the crops which is converted to plant protein to meet our requirement for foods. It has been reported that a stable crop of wheat or cotton may remove 100 kg N/ha.

b) Leaching: Leaching loss of soil-N is by far occurred (>90%) in the form of nitrate which is not adsorbed by the negatively charged soil colloids dominating in most of the soils. The loss of nitrogen in this process depends on application mode of fertilizers, distribution of rainfall, time of irrigation, soil texture, cropping intensity etc. The leaching loss of nitrogen can be minimised by split application of N-fertilizer, nitrification inhibitor, using slow-release N-fertilizer and proper irrigation scheduling.

c) Soil erosion and runoff: Mostly, nitrogenous fertilizer are applied on surface of the soil through broadcasting . The loss of nitrogen through runoff is more when there is heavy rainfall immediately after fertilizer application. Besides, loss of nitrogen through erosion is more under the sloppy areas in a fallow land than under plain land situation.

d) Gaseous losses through NH_3 volatilization or denitrification: The gaseous loss of nitrogen is mainly found as:

   i) Volatilization of ammonia and

   ii) Denitrification

The volatilization loss of ammonia-N from soil occurs when ammonium producing fertilizer is added to an alkaline soil. When, nitrate and nitrite ions are converted into gaseous nitrogenous form is known as denitrification. Under the anaerobic condition, the microorganisms (*Pseudomonas*, *Bacillus*. *Thiobacillus* etc.) can convert NO_3- to N_2 as follows;

$$2NO_3^- \rightarrow 2[O]\ 2NO_2^- \rightarrow 2[O]\ 2NO \rightarrow [O]\ N_2O \rightarrow [O]\ N_2$$

8.4 Losses of Potassium from Soil

a) Leaching loss: The leaching loss of potassium is dominant in coarse textured or organic matter enriched soil of humid region. Split application of potassium is recommended in soils where leaching loss of potassium is high.

b) Soil erosion: The loss of available or total potassium from the soil surface is higher than that of N and P on annual basis.

c) Crop removal: Depending upon the nature of the crops, the average annual loss of potassium due to crop removal varies from 40-400 kg/ha. Sometimes, luxary consumption of potassium is observed by the crops due to uneconomical accumulation of nutrients without corresponding increase in growth or yield of crops.

8.5 Atmospheric Nitrogen Cycle

It has been observed that nitrogen is present in the soil as organic and inorganic forms. When organic nitrogen in the residues of plant and animals undergoes minerelization, then inorganic forms of nitrogen is released.

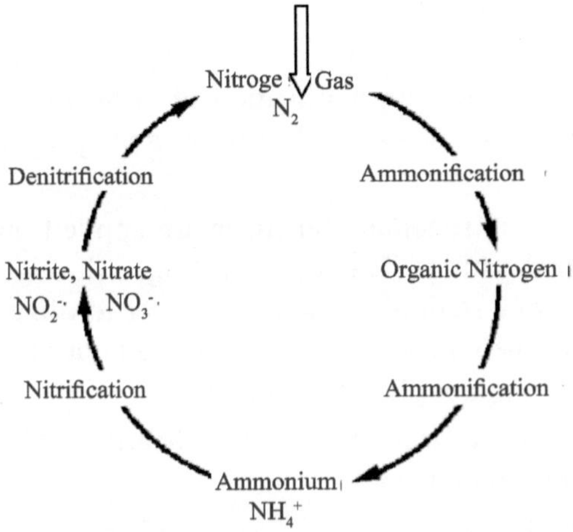

Figure 8.1. Nitrogen cycle

Besides, nitrogen is adsorbed in soil coming from electrical discharge, biological nitrogen fixation and from manures and fertilizers. The produced NH_4^+ is converted into NO_3^- (nitrification) by nitrifying bacteria (Figure 8.1), may be immobilised by soil microorganisms (immobilization), may be taken up by the plants, may be fixed by clay minerals and may be lost to the atmosphere by NH_3 gas (volatilization). The generated NO_3^- on transformation may be taken up by plants, may be leached out by runoff or by drainage water, may be immobilized by microorganisms or may enter into the atmosphere as various nitrogenous gases are produced by denitrification process.

8.6 Potassium Cycle

The potassium is cycled (Figure 8.2) through the soil-plant-animal systems, although, much of the total potassium (> 95%) in soil is within the structure of K-bearing minerals (mica, biotite, feldspar). Due to weathering of minerals, potassium is released into the soil solution directly or by the formation of slowly available non-exchangeable form (NEK) and readily available exchangeable form. The available potassium is taken up by plants from soil or there may be leaching or run off loss. Incorporation of potassium through fertilizer or by decomposition of plant /animal waste may have some bearing on reversion of dynamic equilibrium between the solution potassium and its slowly available form.

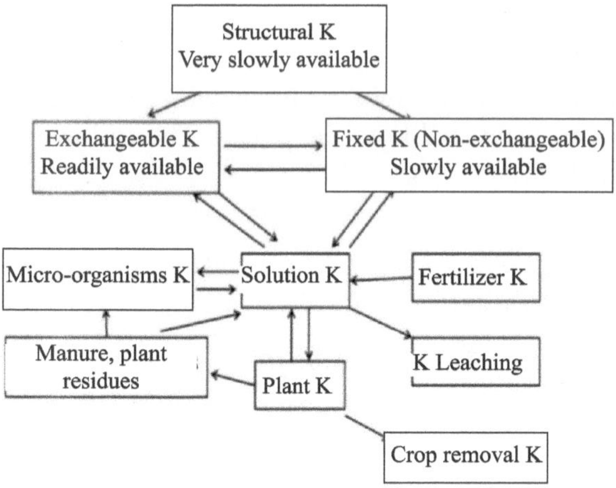

Figure 8.2. Potassium cycle

8.7 Mechanism of Phosphate Fixation in Soil

The inorganic phosphorus may be adsorbed to the surface of the mineral as labile-P or may be precipitated as non-labile P which is known as adsorption and precipitation reaction of inorganic P and is derived from mineralization of organic P, dissolution of P- containing minerals or addition of P-fertilizers. In acid soil, P is adsorbed on the surface of Fe/Al-P, or in $CaCO_3$ or on clay minerals. The P-fixation in soil is associated with the Freundlich and Langmuir adsorption isotherms. At low concentration of P in solution, Freundlich equation fits well, while, the Langmuir equation is reliable at high P concentration in solution and can also predict the maximum sorption capacity of P in soil.

Freundlich equation:

$$m=AC^{1/n} \quad (8.1)$$

Langmuir equation

$$m=ABC/(1+BC) \quad (8.2)$$

Where, m= amount of P sorbed per unit weight of soil or clay; n = constant (n>0).

C= equilibrium concentration of P in soil solution

A and B = constants.

B is related to binding energy.

The P-fixation in soil may be considered under the following aspects:

1. P-fixation in acid soil
2. P-fixation in alkaline soil

8.7.1 Phosphate Fixation in Acid Soils

In acid soils, dominance of iron and aluminium ions are found which may react with the phosphate ions to become unavailable to the plants. The following mechanisms for fixation of phosphates are observed:

Mechanisms of phosphate fixation :

a) Precipitation reaction
b) Adsorption processes

The followings are the different processes of adsorption of P :

i) Precipitation of P from the soil solution: The process is most dominated in the mineral soils where there is presence of Al and Fe in solution and in exchangeable forms over the organic soils. The insoluble hydroxy phosphate is formed on reaction with $H_2PO_4^-$ with Fe/Al ions.

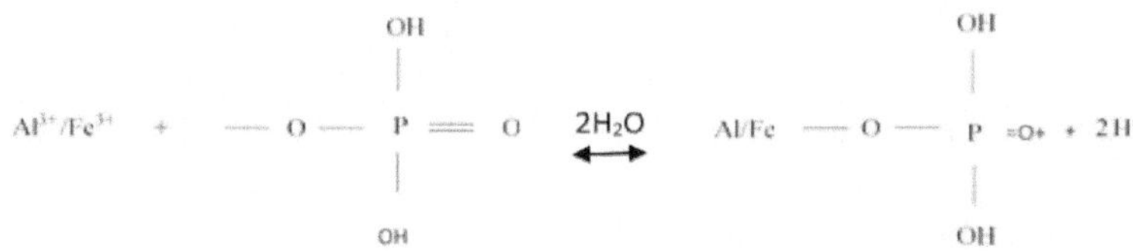

(Soluble - P) (Insoluble - P)

ii) Adsorption of phosphate on silicate clays and oxides/hydroxides of Fe/Al: There is initial attraction of negatively charged phosphate ions to the positively charged hydrous oxides of Fe/Al and silicates and subsequent substitution of structural OH group from the minerals. In other case, there is replacement of adsorbed anions from the negatively charged surface of silicate clays and hydrous oxide of Fe/Al by phosphate ion. With the progress of time, the bond so formed between the P and Al-O/Fe-O produces stable structure (non–labile P) and further lead to the precipitation of hydrous oxides of Fe or Al on this non-labile P-compound as occluded P.

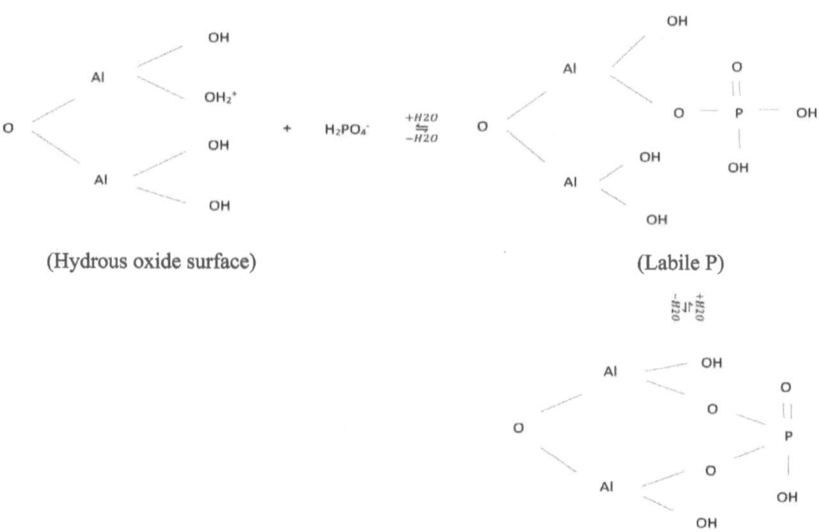

(Non-labile P)

8.7.2 Phosphate Fixation in Alkaline/Calcareous Soils

When the pH of the soil is >8.0, the activity of Ca^{2+} and Mg^{2+} ions are dominant, which on reaction with phosphates may form insoluble compounds like $Ca_3(PO_4)_2$ and $Mg_3(PO_4)_2$ with decreasing solubility.

$$Ca(H_2PO_4)_2.H_2O + 3H_2O \xrightarrow[+Ca^{+2}]{-2H^+} 2(CaHPO_4.2H_2O)\Downarrow \xrightarrow[+Ca^{+2}]{-2H^+} Ca_3(PO_4)_2 + 4H_2O$$

At low P concentration in soil, the soluble P is adsorbed on the surface of $CaCO_3$ by replacing CO_3^{-2} and at high P concentration, the Ca-P minerals precipitate on the $CaCO_3$ surface in the calcareous soil. The insoluble tricalcium phosphate may be transformed into more insoluble form of apatite which in term may be a source of plant available P, if it is applied in acidic soil.

8.8 Factors Affecting Phosphate Fixation

The phosphate fixation in soil is affected by many factors which are given below

i) Nature and amount of clay minerals: The phosphate ions are adsorbed on the surface of the hydrous oxides and hydroxides of iron and aluminium. The less crystalline structure of iron and aluminium compounds having greater surface area can fix more phosphate ion.

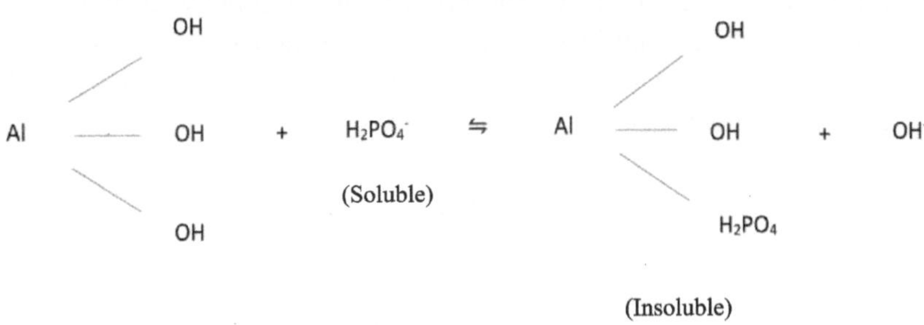

The 1:1 type of clay with the associated hydrated oxides of Fe and Al can retain more phosphate than 2:1 type of clay. The higher amount of clay fraction in soil will fix more phosphorus than lower amount of clay. The amorphous mineral (Allophane) having negative charge when partly/fully balanced by complex aluminium cations can fix phosphorus on reaction with the Al. The presence of $CaCO_3$ can fix phosphate ions due to the less soluble formation of tricalcium phosphate or carbonate apatite at the later stage of reaction. In some cases, calcium act as a linkage between the phosphate and soil colloid and thus helps the fixation processes in soils.

ii) Soil pH: At low pH of the soil, 'P' is fixed more than the alkaline pH (Figure 8.3) and P -availability is more near pH 6.5. With the gradual increase of soil pH, the P-fixation is decreased due to the lowering of the activity of Fe and Al ions, while at pH more than 7.0, the dominant Ca^{2+} and Mg^{2+} and their carbonates can fix P in soil. The overliming decreases P-availability due to the formation of insoluble Ca-P/Mg-P. The order of P-fixation is : Acid soils> Calcareous soil> Alkaline soils.

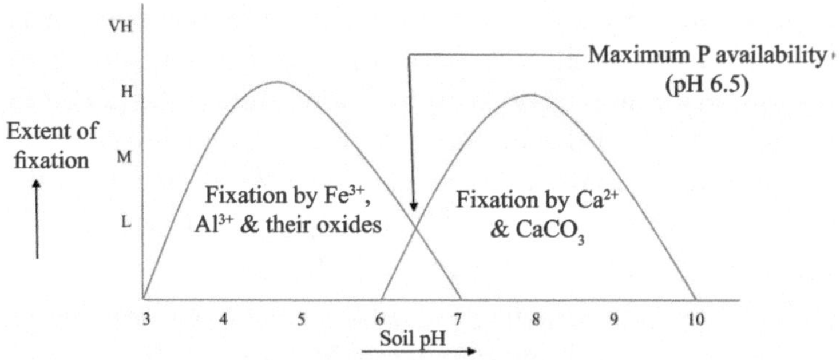

Figure 8.3. Phosphate fixation in soil at different soil pH.

L= low ; M= Medium; H= High ; VH = Very high

iii) Cations and anions: The higher cationic valences can fix more P than the lower valence one due to the greater accessibility of positive charges developed at the edge of clay minerals to phosphate ions.

The phosphate fixation is reduced by the anions originating from the inorganic and organic ions for the competition with P for the same adsorption sites. The NO_3^- and Cl^- have less effect on P-fixation than OH^-, SO_4^{2-}, MoO_4^{2-} and $H_3SiO_4^-$. The replacement of the adsorbed phosphate by the other competing anions depend on the bonding strength of the ions with the adsorption sites. The organic anions like citrate, oxalate, tartarate, malate etc. when form complex with Fe^{3+}/Al^{3+} do not react with $H_2PO_4^-$. The negatively charged organic anions also compete with $H_2PO_4^-$ for the same adsorption sites and reduce P-fixation in soils. Sometimes, the humic molecule resist the P-ion to react with the exchange complex by surface coating on the reacting sites and thus reduce the fixation of P in soils.

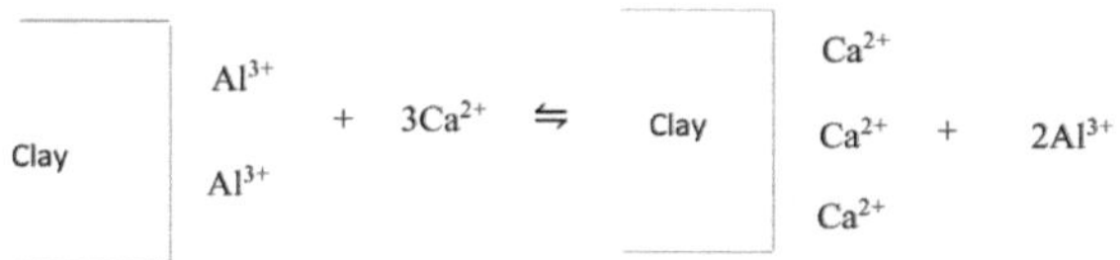

$$Al^{3+} + 2H_2O \longleftrightarrow Al(OH)_2^+ + 2H^+$$

$$Al(OH)_2^+ + H_2PO_4^- \longleftrightarrow Al(OH)_2H_2PO_4$$

iv) Redox reaction: The oxidation-reduction (redox) reaction in soil under submergence can convert the insoluble ferric phosphate to more soluble ferrous phosphate and in calcareous soil, the solubility of Ca-P is increased.

8.9 Phosphorus Cycle

The dynamic equilibrium of P in soil (Figure 8.4) is correlated with the P-uptake by the plants and replenishing the P back into the soil. The soil solution -P may be taken up by the plant roots , by soil microorganisms (immobilization); may be adsorbed in the minerals or on the clay surfaces as labile inorganic –P or may be lost through surface runoff , soil erosion and leaching. The deficit of P in soil may be replenished by the addition of fertilizers, minerelization of organic residues , desorption of sorbed-P from clay mineral surface or by the dissolution of P from non-labile primary and secondary P minerals.

Soil solution P ⟺ Labile P ⟺ Non-labile P

The dynamics of P will help understanding the need for judicious application of P- fertililizer in soil for the greater growth and yield of crops.

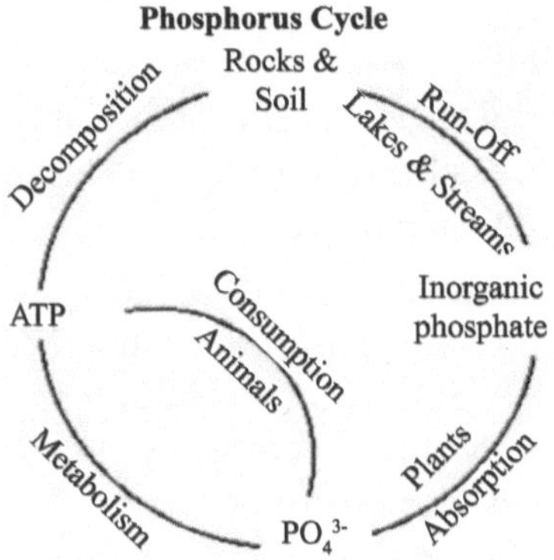

Figure 8.4. Phosphorus cycle

8.10 Losses of Phosphorus from Soil

The losses of P from soil may occur by: a) crop removal b) leaching c) erosion of soil.

a) Crop removal: The type of crop and its targeted yield can influence P-removal from soil at harvest.

   e g. Wheat- 20 kg/ha, rice- 21 kg/ha, Maize- 42 kg/ha.

b) Leaching loss: The soluble-P when converted into insoluble- P compound becomes less mobile from the point of application to the rhizosphere zone. The loss incurred by this process is relatively less significant compared to the leaching loss of N_2 from the soil.

c) Erosion of soil: This may happen when there is surface erosion of soil and exposing the subsurface soil from the cropped field.

Questions

Objective

1. $x/m = kc^{1/n}$ is
   a) Langmuir's equation b) Freundlich equation
   c) Gapon's equation d) Vanselaw's equation
2. At the pH range of 2-5, phosphorus is mainly fixed by
   a) Fe & Al b) Clay
   c) Ca & Mg d) Organic matter
3. The reason for soil flocculation is
   a) Lowering of zeta potential b) Increase of zeta potential
   c) No effect of zeta potential d) None is true
4. Phosphate fixation is prominent in
   a) Acid soil b) Alkaline soil
   c) Saline soil d) Salaine-alkaline soil
5. Carbon dioxide content in soil is
   a) 0.3% b) 0.03%
   c) 3.0% d) None of the above
6. The "Lattice hole theory" was proposed by
   a) Page and Baver b) Kate and Nelson
   c) Baver and Sommner d) Bray and Kurtz
7. Langmuir equation:
   a) m=ABC/(1+BC) b) m=AB/(1+BC)
   c) m=AC/(1+BC) d) m=A/(1+BC)
8. $Al^{3+} + 2H_2O =$
   a) $Al(OH)_2^+ + 2H^+$ b) $Al(OH)_2^+ + 3H^+$
   c) $Al(OH)_2^+ + 1/2H^+$ d) $Al(OH)_2^+ + 4H^+$
9. Insoluble compounds like $Ca_3(PO_4)_2$ and $Mg_3(PO_4)_2$ with decreasing solubility are found in:
   a) Acid soil b) Alkali soil
   c) Neutral soil d) Sodic soil

10. The redox reaction in soil under submergence can convert the
    a) Ferric phosphate to ferrous phosphate
    b) Ferrous phosphate to ferric phosphate
    c) Zinc phosphate to zinc
    d) None of the above

Subjective

1. What do you mean by fixation of nutrients in soil ?
2. What is the effect of pH on ion fixation in soil ?
3. What are the different fractions of soil K ?
4. What is the role of phosphate solubilizing bacteria in P release from soil.
5. What is volatilization loss of nitrogen ?
6. Why phosphate fixation is more in acid soil than alkaline soil ?
7. How fixed K can be released from soil ?
8. What are the adsorption equations for P-fixation in soil.
9. What is the dynamic equilibrium of P in soil ?
10. How leaching loss of nitrogen in soil may be checked ?

Descriptive

1. Outline in brief the fixation of NH_4^+ or K^+ in soils.
2. Describe the mechanism of P-fixation in soil.
3. What are the factors affecting NH_4^+ or K^+ in soils.
4. What are the factors affecting losses of Nitrogen or Potassium in soil ?
5. Briefly describe the Nitrogen or Potassium cycle in soil.
6. What are the factors affecting P-fixation in soil ?
7. Describe the p-cycle and the factors affecting P-loss from soil.
8. What are the empirical models governing the P-fixation in soil ?
9. What is the agricultural significance of nitrogen fixation in soil ?
10. What are the different management options for improving the use efficiency of N,P and K in soil ?

Suggested Readings

Havlin, J. L., Tisdale, S.L., Beaton, J.D. and Nelson, W.L. (2007). Soil fertility and fertilizers. (Seventh Edition). First impression. Prentice Hall of India Ltd. New Delhi.

Lindsay, W.L. (1979). Chemical Equilibria in Soils. New York; John Willey and Sons.

Pal, S.K. (2016). Text book of Soil Science. Oxford and IBH Publishing Company Pvt. Ltd.

9

Chemistry and Implications of Submerged Soil

Understanding the chemistry of submerged soil in the field of Geology, Pedology, Agriculture and other allied sectors draws attention to the research on characteristic changes in soil under submergence. It has been reported that, about 72% of the earth's surface is covered by submerged soils or sediments . Rice plants have unique adaptation capability to grow under submerged soil and in any type of soil (acid/alkaline) condition or land situation (upland/lowland). The knowledge on changes of important physical, chemical and biological properties under submergence have important role to play in understanding the mechanisms of availability of nutrients to the plants and sustainability of the crop during its growing period.

9.0 Submerged Soil

The soils which are saturated with water for a longer period of time in a year to make the soil a distinctive gley horizons resulting from oxidation-reduction processes are called submerged soil. About 72% of the earth's surface is covered with soil –submergence or sediments.

The submerged soil may show the following features

i) A partially oxidized 'A' horizon high in organic matter.

ii) A mottled zone in which oxidation and reduction processes alternate

iii) A permanently reduced zone which is bluish green in colour due to the presence of ferrous compounds like magnetite, pyrite, ferrous silicate etc.

The soil is intermittently saturated with water, oxidation of organic matter is relatively slow and accumulation is mostly in the "A" horizon. In the second layer, the mottles of Fe and Mn are deposited or streaks are found , if the diffusion of O_2 into the soil is slow, while concretions are observed if the diffusion is rapid in the system.

9.1 Types of submerged soil

There are different types of submerged soils based on duration of submergence, types of minerals and organic matter.

9.1.1 Waterlogged or Gley Soil

A soil which are annually saturated with water for a longer period of time bearing a distinctive gley horizon for the redox reactions and whose properties are dominated by this process is called a gley (UK), gleysol (World Reference Soils Base) or ultisol (USDA). Due to the submersion of the sub-soil below the water table (a ground water gley), this type of soil can develop in low-lying areas such as floodplains , where drainage is poor because of the higher elevation of the groundwater table, although, surface-water *gleying* may occur when precipitation input at the surface does not drain freely through the groundwater.

9.1.2 Marshy Soil

Marshy soil is rich in iron with varying amount of organic matter. This type of soil is found in the coastal areas of Orissa, Tamil Nadu, West Bengal, some parts of Bihar, and in very few areas of Uttarakhand. This soil develops itself due to stagnation of water having high water retention capacity. The Peat and Marshy soils are found in Kerala, coastal track of Orissa, Sunderban area of West Bengal. When the vegetation growing in such wet places dies, it decomposes very slowly dues to excessive wetness of soils and after several hundreds of year a layer of partly decayed organic matter accumulates on the surface, giving rise to such peaty and marshy soils which are black coloured, heavy and highly acidic in reactions. The soils produce good crops of rice when well drained and properly fertilized. These soils are more or less permanently saturated or submerged. The Upland marshes receive mainly rainwater and are poor in bases with the pH values of 3.5- 4.5, while lowland marshes are saturated or submerged with water-carrying bases and have pH varies from 5.0-6.0.

9.1.3 Paddy Soil

Paddy soils are used for the cultivation of rice. Mostly the paddy soils are usually Entisols and Inceptisols and occur in landscapes where both surface and internal drainage are poor. The paddy soils are found in deltas and adjacent flood plains, in valleys, and coastal plains. Paddy soil is irrigated during the growth period of rice to meet the consumptive use of water. It requires to maintain 5-10 cm. of standing water during the 4-5 months growing period of the crop and then draining and drying the fields at harvest of the crop. Due to the prolonged submergence , the paddy soil comes to a neutral pH.

9.1.4 Subaqueous Soils

Under the continuous water columns, the soils are formed from river lakes and ocean sediments. In general, the uppermost layers of unconsolidated river, lake and ocean sediments are considered as subaquatic soils.

9.2 Characteristics of Submerged Soils

Different physico-chemical changes are found in soils under prolonged submergence with water. The biotic and abiotic ecosystem are also affected due to the stagnation of water on soil. The following changes are observed due to the soil submergence.

9.2.1 Depletion of Oxygen

Due to the submergence of soil, air in the pore spaces are replaced by water and most of the soil layers except in a thin layer at the soil surface, and sometimes a layer below the plough sole, virtually become oxygen-free within a few hours after submergence. The oxygen-diffusion in the stagnated layer above the soil becomes slow and the rate of O_2 consumption is reduced to a considerable level. As a result of that, the soil under submergence becomes practically devoid of oxygen and still the oxygen trapped in blocked pore spaces is rapidly utilized by facultative anaerobic organisms, which leads to the formation of two different layers: an oxidized or aerobic surface layer where O_2 is present and a reduced or anaerobic layer in which no free O_2 is present.The thickness of these two layers depend upon the supply and consumption of oxygen in these layers. The ratio of O_2 supplied from the atmosphere to the O_2 consumption in the soil could determine the thickness of the oxygenated surface layer in the soil. During the process of aerobic respiration, the final electron acceptor is oxygen which on reaction with H^+ ions form water.

$$O_2 + 4H^+ + 4e^- \rightarrow 2H_2O \qquad (9.1)$$

In absence of free oxygen , a number of oxygen containing compounds will accept electrons and lose their oxygen.

$$NO_3^- + 2H^+ + 2e^- \rightarrow NO_2^- + H_2O \qquad (9.2)$$

$$NO_2^- + 8H^+ + 6e^- \rightarrow N_2 + 4H_2O \qquad (9.3)$$

Some higher valence cations or hydrogen ion can act as electron acceptor and are reduced to their lower valence state or hydrogen gas respectively.

$$Fe^{+3} + e \rightarrow Fe^{+2} \qquad (9.4)$$

$$2H^+ + e \rightarrow H_2 \qquad (9.5)$$

9.2.2 Accumulation of Carbondioxide

Accumulation of carbon dioxide and methane developed under the submerged soil may escape as bubbles if sufficient pressure are developed in the soil. The partial pressure of CO_2 (pCO_2)will affect the solubility of the cations Ca^{2+}, Mg^{2+}, Fe^{2+} and Mn^{2+} as well as its pH, Eh and exchange reactions. There may be decrease in PCO_2 after one to four weeks of submergence. There is bacterial reduction of CO_2 to CH_4 and CO_2 as insoluble carbonates may be removed from soil.

The dissolved CO_2 in the soil solution can affect chemical equilibria or the solubility of cations which are summarised below;

$$pH + K - \frac{2}{3}\log pCO_2 \qquad (M\ CO_3\ present)\ (9.6)$$

$$pH + \frac{1}{2}\log M^{2+} = K^1 - \frac{1}{2}\log pCO_2 \qquad (M\ CO_3\ present)\ (9.7)$$

$$pH = K - \frac{1}{3}\log pCO_2 \qquad (M\ OH_2\ present)\ (9.8)$$

$$Eh = Eo + 2.303\frac{\eta^n RT}{\eta F}\log pCO_2 - 2.303\frac{\eta^n RT}{\eta F}pH \qquad (M\ CO_2\ present)\ (9.9)$$

The higher pCO_2 in acid soils than in neutral soil can be explained by the following equation (Eq.9.10)

$$K = (H^+)(HCO_3^-)/(H_2CO_3) \qquad (9.10)$$

$[H_2CO_3 = H^+ + HCO_3^-]$

The pCO_2 will increase with the increase in the activity of H^+ ions at a particular temperature so that the equilibrium constant(K) is maintained.

9.2.3 Soil Reduction

Due to the submergence of soil , the redox potential (Eh,) of soil is decreased which leads to oxygen depletion followed by a chain of soil chemical changes (Table 9.1). The reduction of the soil is due to the anaerobic respiration by the soil microorganisms. The soil rich in organic matter but poor in Fe and Mn content maintain high $pCO_{2.}$ Under this situation, transplanting of rice should be delayed by at least two weeks after flooding to minimize CO_2 injury.

It is found that , in any spontaneous chemical reactions, the free energy (ΔG) is lowered down . From the equation,

$Ox + ne \leftrightarrow Red$

$$\Delta G = \Delta G^0 + RT\ln (Red)/(Ox) \qquad (9.11)$$

where, Ox and Red are the activities of oxidised and reduced species (Eq. 9.11) respectively and ΔG^0 is the standard free energy change when the activities of oxidised and reduced species are at unity. Again ,the decrease in free energy of the process is equal to the electrical energy obtained from the equations.

i.e., $\Delta G = -nEF$ (electrical energy produced)

where, n = number of equivalents of reaction occurring,

F = Faraday constant

E = EMF of the cell.

According to the Nernst equation,

$E = E_0 + RT/nF \ln (Ox/Red)$ (9.12)

where,

E= potential of the reduction

E_0 = potential when the activity of the oxidised and reduced species are at unity

F= Faraday constant (96500 Coulombs/mol)

R= Gas constant (8.314 J/mol/K)

T= Absolute temperature

n= change in oxidation number. E the standard hydrogen - electrode , is denoted by Eh and the equation will be transformed into :

$$Eh = E_0 + RT/nF \ln (Ox/Red) \quad (9.16)$$

Eh is positive in strongly oxidised system and is negative and is negative and low in strongly reduced system.

9.2.4 Electrochemical Changes

The main electrochemical changes under submerged soil that influences the growth of rice are:

i) Decrease in redox potential

ii) Increase in pH of acid soil and decrease in pH of alkaline soils

iii) Increase in specific conductance

iv) Cation exchange reactions involving Fe^{2+}

v) Sorption and desorption of ions

The reduction of soil takes place when a soil is submerged and the reduced soil contains reduced counter parts of NO_3^-, SO_4^{2-}, Mn^{4+}, Fe^{3+} and CO_2 as NH_4^+, H_2S, Mn^{2+}, Fe^{2+} and CH_4. The reduction of different soil components takes place in a sequential form; nitrates are reduced first followed by manganic (Mn^{4+}) and so on. Oxidation-reduction is a chemical reaction in which electrons are transferred from a donor to an acceptor. The electron donor losses electrons and increases its oxidation number or is oxidised, the acceptor gains electrons and decreases its oxidation number or is reduced. The source of electrons for biological reductions is the organic matter.

The driving force of a chemical reaction is the tendency of the free energy of the system to decrease until, at equilibrium, the sum of the free energies of the products equals that of the remaining reactants. In a reversible oxidation - reduction reaction, this force can be measured in calories or in volts. The change in free energy, $\Delta G°$ for the reduction;

$$Ox + ne \rightleftarrows Red \text{ is given by}$$

$$\Delta G = \Delta G^0 + RT\ln (Red)/(Ox) \qquad (9.13)$$

where, (Red) and (Ox) are the activities of the reduced and oxidised species and $\Delta G°$ is the free energy change when the activities are unity. Converting calories to volts using the relationship,

$$\Delta G = -nEF \qquad (9.14)$$

n = number of equivalents of reaction occurring

ΔG = Electrical energy produced, where n is the number of equivalents of reaction occurring.

From the following equation, (Nernst equation)

According to the Nernst equation,

$$E = E_0 + (RT/nF) \ln (Ox/Red) \qquad (9.15)$$

in which E is the voltage of the reaction, E_0 is the voltage when (Ox) and (Red) are each unity, and F is the Faraday constant in heat units. If E is measured against the standard hydrogen electrode, it is denoted by Eh and Equation 9.15 becomes.

$$Eh = E_0 + (RT/nF) \ln (Ox/Red) \qquad (9.16)$$

Eh is a quantitative measure of the tendency of a given system to oxidize or reduce substances. The value of Eh is positive (+ve) and high in strongly oxidising systems and is negative (-ve) and low (Figure 9.1) in strongly reducing systems. The chemical reaction which involves the exchange of electrons will be influenced by redox potential (Eh).

When a soil (aerobic) is submerged, a sharp drop in the potential will be observed and within a few weeks negative (-ve) potentials will be developed in most of the soils. The development of Eh may be as low as -300 to -400 mV depending upon presence of soil organic matter, Fe and Mn content, pH and bacterial potentials.

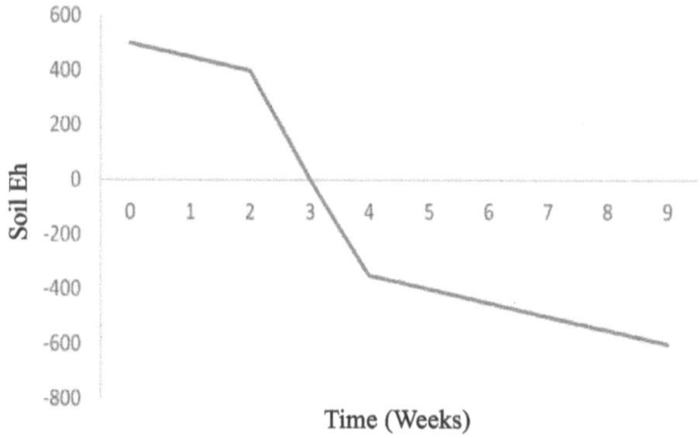

Figure 9.1. Changes in Eh on submergence of soil

The value of Eh is important considering the growth, nutrition and yield of rice to understand the situations and its proper management.

9.2.5 Highly Oxidised State

Initially in uplands, the NH_4^+ -N may be oxidised to NO_3^- - N and may leach or denitrify. The oxidation of organic residues leads to the liberation of CO_2, NO_3, and SO_4 etc. The availability of iron, manganese and phosphorus will be very limited since they are present as their oxidised forms (Fe^{3+} and Mn^{4+}) and phosphorus as insoluble iron and manganese phosphates. These may lower down the soil pH which can lead to aluminium toxicity in crops.

9.2.6 Healthy Reduced State

The NH_4^+ -N is present in large amounts and is absorbed on soil colloids resulting very little leaching and denitrification losses. The availability of iron will increase due to reduction of Fe^{3+} to Fe^{2+}. Generally H_2S will not form at this stage, but if it forms, it will precipitate as FeS (Eq. 9.17)

$$\begin{aligned} Fe^{3+} + e^- &\rightleftharpoons Fe^{2+} \\ Fe^{2+} + H_2S &\longrightarrow \underset{\text{(insoluble)}}{FeS\downarrow} \end{aligned} \tag{9.17}$$

The pH comes around 6.5 which is optimum for the availability of all the macro- and micro-nutrients.

9.2.7 Highly Reduced Toxic State

There may be appearance of iron toxicity at highly reduced condition of the soil because of its maximum concentration. Sulphide (S^{2-}) ions may accumulate and cause sulphide toxicity to rice roots. The generation of organic acids (butyric acid, propionic acid etc.), ethylene, mercaptans, organic sulphides are found which could create problems to the rice plant (Figure 9.2).

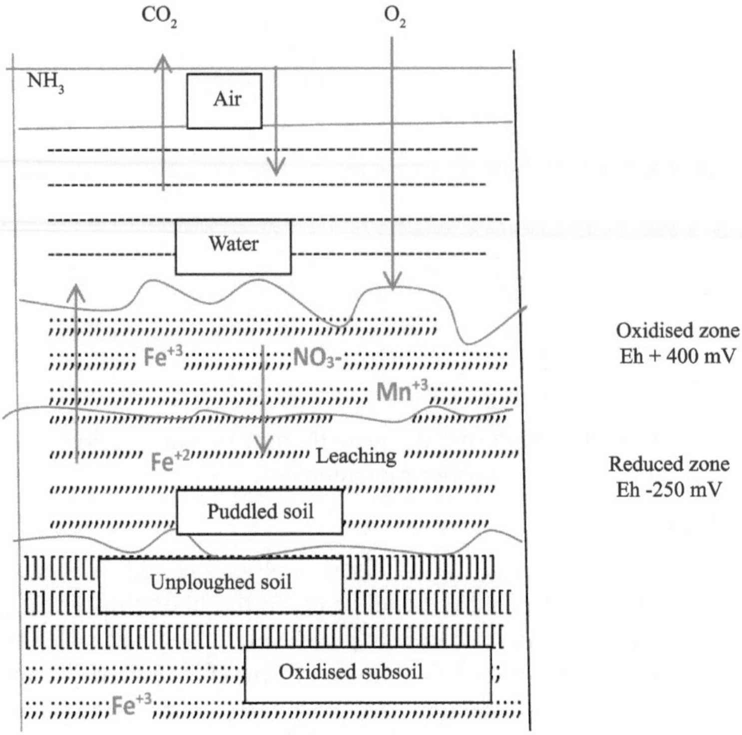

Figure 9.2. Characteristics of oxidised and reduced zones in submerged soil

i) The rice plants may be affected due to the low Eh in soil submergence. The germination and seedling emergence may be affected at low Eh but not growth of the well-established plants.

ii) Low Eh destroys NO_3^- - N but favours NH_4^+ - N accumulation and nitrogen fixation and hence results net increase in the soil nitrogen stock.

iii) Low redox potential benefits rice by increasing the availability of N, P, Si, Fe, Mn, and Mo.

iv) The availability of S, Cu and Zn is decreased for rice.

9.3 Sequential Reduction of Oxidation-Reduction Systems in Submerged Soils

Due to depletion of O_2 in submerged soils, the need for electron acceptors by facultative and obligate anaerobic micro-organisms results in the reduction of several oxidised soil components. Microbes utilize the available oxidised soil components and reduce the oxidation number of the oxidised atom. The sequential reduction of elements (Table 9.1) under soil submergence are given hereunder.

Table 9.1. Sequential reduction of the oxidised soil components under submergence

System	*Redox potential (E_o), mV*
$O_2 + 4H^+ + 4e^- \rightleftharpoons 2H_2$	+830
$NO_3^- + H_2O + 2e^- \rightleftharpoons NO_2^- + 2OH^-$	+430
$MnO_2 + 4H^+ + 2e^- \rightleftharpoons Mn^{2+} + 2H_2O$	+410
$Fe(OH)_3 + e^- \rightleftharpoons Fe^{2+} + 3OH^-$	+130
$CH_3COCOOH + 2H^+ + 2e^- \rightleftharpoons CH_3CHOH + COOH$	-180
$SO_4^{2-} + H_2O + 2e^- \rightleftharpoons SO_3^{2-} + 2OH^-$	-490
$CO_2 + 2H + 2e^- \rightleftharpoons HCOOH$	-620
$H_3PO_4 + 2H^+ + 2e^- \rightleftharpoons H_3PO_3 + H_2O$	-700
$H_3PO_3 + 2H^+ + 2e^- \rightleftharpoons H_3PO_2 + H_2O$	-920
$H_3PO_2 + H^+ + e^- \rightleftharpoons P + 2H_2O$	-930

The general sequence of reduction (Figure 9.3) with the days of submergence indicates the changes of the oxidised state of the elements into reduced state.

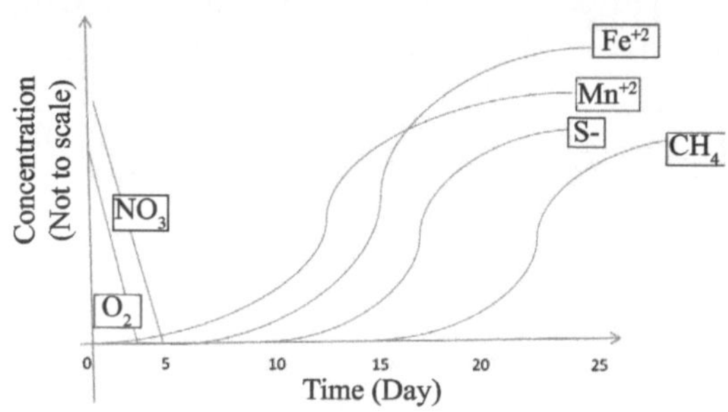

Figure 9.3. Sequential reduction of various soil redox systems in submerged soils

The approximate critical redox potential (Table 9.2) at which these systems become unstable is useful for characterizing the intensity of anaerobiosis in soils.

Table 9.2 Some important oxidised components with redox potential values in submerged soil

Reduction	Redox potential (mV)
$O_2 \rightarrow H_2O$	+380 to +320
$NO_3 \rightarrow N_2$; $Mn^{+4} \rightarrow Mn^{+2}$	+280 to +220
$Fe^{+3} \rightarrow Fe^{+2}$	+180 to +150
$SO_4^{2-} \rightarrow S^{-2}$	-120 to -180
$CO_2 \rightarrow CH_4$	-200 to -280

9.3.1 Concept of pE

The pE is the measurement of electron activity which is used in some cases instead of Eh while studying redox equilibria. Just as pH is a measure of proton (H^+) activity, so in pE, the negative logarithm of the electron activity, a measure of electron activity. It can be written as,

$$pE = -\log (e) \quad (9.18)$$

$= Eh/2.303\ RTF^{-1}$

or, $pE = Eh/0.0591$

$$pE_0 = E_0/0.0591 \text{ at } 25°C \quad (9.19)$$

So, for the determination of pE, the value of redox potential (Eh) is divided by 0.0591 . In strongly oxidising system the electron activity is low and pE is large and positive while in reducing system, pE is small and negative. Here, F = Faraday constant.

9.3.2 Soil pH

When an aerobic soil is submerged, its pH decreases during the first few days reaches a minimum, and then decreases asymptotically to a fairly stable value of 6.7-7.2 after few weeks. The overall effect of submergence is to increase the pH of acid soils and to decrease the pH of alkaline soils. Thus, submergence makes the pH values of acid soils (except those low in iron) and alkaline soils converge to neutral soil pH 7.0 (Figure 9.4).

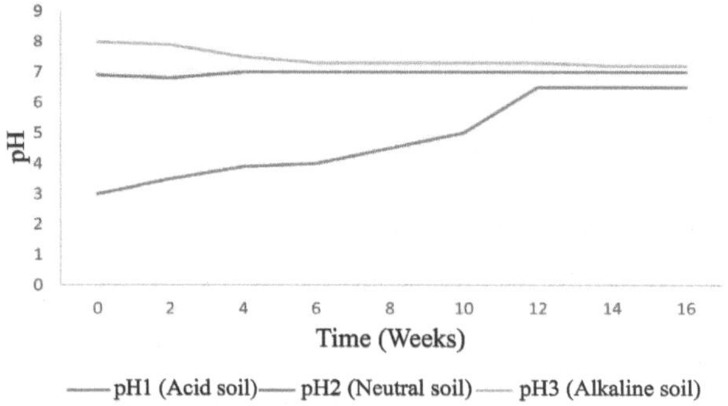

Figure 9.4. Effect of submergence on soil pH

The increase in pH is acid soils may be due to the reduction of iron (Fe^{3+}) and manganese (Mn^{4+}) and also due to the presence of organic matter that enhance the process of soil reduction. Since the reduction process removes hydrogen ions (H^+) from the solution with the progress of the reduction, hence, the pH of the acid soil rises. Soils, however, contain CO_2 dissolved in the soil solution, and if the conditions are suitable, either ferrous carbonate ($FeCO_3$) will be precipitated in the soil solution, through its effect on pH, will modify the concentration of ferrous ions (Fe^{2+}) in equilibrium will be transformed into ferroso-ferric hydroxides [$Fe_3(OH)_8$]. The decrease in pH of alkali and calcareous soils may be due to the release of sufficient amount of CO_2 during decomposition of organic matter. The increased concentration of CO_2 produces hydrogen ions (H^+) in the system thereby decreases the soil pH.

$$\text{Organic matter} \xrightarrow{\text{Decomposition}} CO_2 \xrightarrow{H_2O} H_2CO_3 \xrightarrow{\text{Dissociation}} H^+ + HCO_3^-$$

again,

pH = 6.03-0.67 log pCO$_2$ (9.20)

The pH of submerged alkali and calcareous soils under submergence are generally governed by the Na_2CO_3 -H_2O-CO_2 and $CaCO_3$- H_2O-CO_2 system respectively.

pH= 7.85 + log [Na$^+$] – log pCO$_2$ -0.511$^{1/2}$ (9.21)

Soils high in organic matter and reducible iron attain a pH of about 6.5 within a few weeks of submergence. Acid soils low in organic matter or active iron slowly attain pH values which are less than 6.5. But the acid sulphate soils, low in iron may not attain a pH of more than 5 even after months of submergence. Organic matter magnifies the decrease in pH of sodic and calcareous soils. Lowering of temperature or the presence of nitrate (NO_3^-) retards the increase in pH.

9.3.3 Specific Conductance

The specific conductance of the solution of most soils increases on submergence, attain a maximum and declines to a fairly stable value, which varies with the nature and properties of the soil (Figure 9.5).

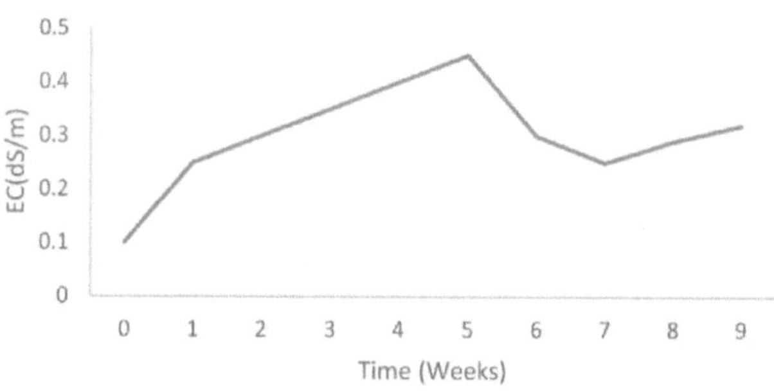

Figure 9.5. Changes in specific conductance due to submergence

The release of Fe^{2+} and Mn^{2+} from the insoluble Fe^{3+} and Mn^{4+} oxide hydrates, the accumulation of NH_4^+, HCO_3^- and $RCOO^-$, and in case of calcareous soils, the dissolution of $CaCO_3$, $MgCO_3$ by CO_2 and organic acids could lead to the increase in specific conductance during the initial few weeks of submergence.

There may be increase in the concentration of calcium and magnesium by the displacement of ions, from the soil colloids.

$$Ca^{2+}\boxed{Clay} + Fe^{2+} \rightleftharpoons Fe^{2+}\boxed{Clay} + Ca^{2+}$$

(soil solution)

$$\Downarrow$$

$Fe_3O_4.nH_2O$ (precipitation)

(Insoluble hydrated ferric oxide)

There is decline of the conductance in the soil solution after the maximum increase due to the precipitation of Fe^{2+} as $Fe_3O_4nH_2O$ and Mn^{2+} as $MnCO_3$. In calcareous soils, the decrease is caused by the fall in partial pressure of $CO_2(pCO_2)$ and the decomposition of organic acids. In the strongly acid lateritic soils, a steep increase in specific conductance is observed during first week of submergence and the soils high in organic matter could attain higher peaks.The specific conductance increases gradually to a level of 1-2 mmhos cm^{-1} in a soil where rate of reduction is slow and thereafter declines slowly. The calcareous soil low in organic matter shows very little change in specific conductance in the soil-water systems.

9.3.4 Ferrolysis

It is a process of decomposition of clay and transformation of iron oxides to ferrous (Fe^{+2}) ions under the influence of periodic reduction. Ferrolysis comprises of processes under various components; some operating in reduced conditions (Wet season) alternating with other operating in oxidising condition (Dry season). The schematic representation of Ferrolysis during wet and dry season are summarised below.

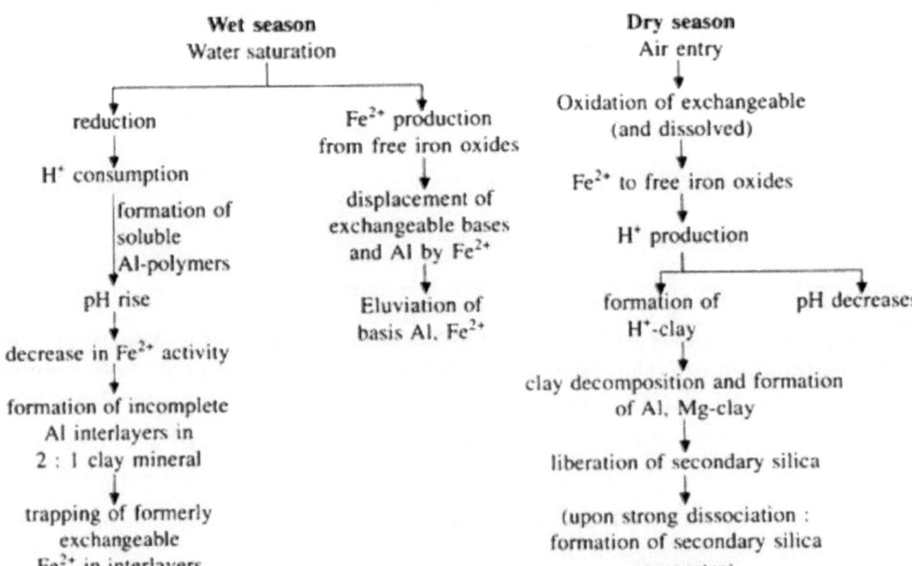

Ferrolysis is not desirable for rice cultivation. There is possibility of clay decomposition in the first few days after air-entry and slower distribution of bases. Ferrolysis can be controlled by the following ways :

Control of Ferrolysis

i) By eliminating the oxidation phase.

ii) By eliminating the reduction phase. This can be done by rapid leaching which would encompasses great losses of nitrogen and other nutrients or by shifting to upland rice.

iii) Raising the dry season (oxidised) pH by liming.

iv) By replacing the basic cations either through fertilization or liming or base-rich irrigation water.

Chemical transformation of nutrients

Both the macro and micro nutrients undergo transformations due to the submergence of soil.

Nitrogen: In submerged soil , the availability of nitrogen is a challenge because of its transformation and subsequent loss from soil. Here, the mineralization of N is restricted to ammonification stage. Ammonification of organic matter also proceeds at a low rate and some intermediate products of decomposition are obtained.

$$N_2\ N_2\ \text{(Denitrification)}$$

$$\text{(N-fixation)}\ O_2 \uparrow O_2$$

$$\text{Protein} \rightarrow \text{Amino acids} \rightarrow NH_4^+ \rightarrow NO_2^- \rightarrow NO_3^-$$

$$\leftarrow \text{Ammonification} \rightarrow \leftarrow \text{Nitrification} \rightarrow$$

Mineralization: Under the submerged soil, the minerelization of organic nitrogen restricted mostly to ammonification stage due to the want of oxygen required for the subsequent nitrification. But, in the soil-water interface, (aerobic), the nitrate minerelization may end upto nitrate production. The rate of ammonification of organic substances becomes slower due to less efficient anaerobic bacteria. Due to the incomplete breakdown of the carbohydrate, the decomposed product like methane, organic acids (formic, acetic, propionic,butyric etc.) are accumulated. Under the anaerobic situation the nitrogen factor (the number of grams of additional inorganic nitrogen required to decompose 100g of substrate) is lower than aerobic decomposition. The substrate with wide C:N may depress its (N_2) availability in soil having low organic matter. If the temperature raises from 15 to 40^{0}C, ammonia production from the anaerobic soil becomes higher.

Denitrification: The nitrate under anaerobic situation undergoes assimilation and dissimilation (denitrification). The nitrate is incorporated into cell on reduction during assimilation while the nitrate acts as an electron acceptor in place of free O_2 during dissimilation.

On submergence, when the O_2 concentration falls < 0.1 ppm in soil solution, the facultative microorganisms (anaerobic) can transform nitrate to nitrite or some intermediates to nitrogen or its oxides. The rate of denitrification depends on the content of organic matter and on nitrate concentration . It has been observed that, all the nitrates present in soil disappears within a few days of submergence. Application of ammonium fertilizers to the reduced zone under the continuous submergence could minimise the loss through denitrification.

Volatilization: The soils low in CEC, high pH and high soil temperature could favour the ammonia volatilization loss under the soil submergence by the anaerobic decomposition of organic matter.

Leaching loss: A huge amount of nitrate-nitrogen may be leached out from the oxidized zone to the reduced zone in soil (coarse textured, high hydraulic conductivity) during soil submergence.

Nitrogen fixation: The aerobic and anaerobic bacteria play an important role in nitrogen fixation under rice soils. The aerobic bacteria can thrive in the oxidised surface while the anaerobic organisms are in the reduced soil layer. The aerobic bacteria like *Anabaena, Nostoc, Aulosira, Calothrix* can fix about 20-30 $kgha^{-1}$ Nitrogen in soil under slightly alkaline condition . The Azolla-Anabaena association can fix nitrogen in rice soil.

Phosphorus

The important characteristics of P-transformation under the submerged soils are: The mobility of P becomes higher due to the increasing rate of diffusion, although the P mobility is restricted to the top 30 cm layer of the soil.

i) There may be readsorption of P by the silicates / Al-hydroxides.

ii) The solubility of P -compounds may be increased due to the presence of CO_2.

iii) Hydrolysis of P compounds and its solubility will be more (available P).

iv) The mineralization of organic P will be higher on soil submergence.

The phosphorus is present in soil as soluble (water soluble, citric acid soluble) or as solids(Fe/Al/Ca-phosphates) or as organic phosphates. The solubility of Fe/Al phosphate increases with the increasing pH, while that of Ca-phosphate with the decrease in soil pH. It has been observed that, the soil submergence can markedly increases the availability of both native and applied P because of its attachment with number of redox elements rather the transformation is associated with Eh and pH changes in flooded soils.

There is increase in available P on submergence of an acid soil because:

i) Release of P due to higher rate of mineralization of organic residues with increasing pH.

ii) More soluble $Fe_3(PO_4)_2$ is formed on reduction of $FePO_4$ and $AlPO_4$ due to the increase in pH.

iii) Reduction and solubilisation of ferric hydroxide coating which leads to the release of occluded –P.

iv) Release of P from clay surfaces by the anion exchange.

However, there may be decrease in P -solubility after attaining the peak value for the readsorption of soluble P by silicate clays and hydrous oxides of Al in the anaerobic zone, or diffusion to oxidised zone or reprecipitation. Organic matter addition under the soil submergence may cause chelation. Furthermore, there may be decrease in available P due to the transformation of inorganic P to organic form (immobilization).

Potassium

The redox reaction of existing cations in soil-submergence can influence the solubility of K. The reduced species of Fe^{+2} and Mn^{+2} under submergence can replace K^+ from the exchangeable phase to soil solution phase. Decomposition of SOM also can release K^+ either from acid/alkali soils. Where the K-status is low in a specific soil, continuous submergence or alternate wetting and drying could facilitate K^+ release from the non-exchangeable sites.

Sulphur

Under the soil submergence, the sulphate form of sulphur is reduced to sulphide and there is breakdown of S-containing aminoacids (cysteine, cysteine and methionine) to H_2S which can react with Fe^{+2} to yield FeS

$$SO_4^{2-} \rightarrow H_2S$$

$$Fe^{+3} \rightarrow Fe^{+2}$$

$$Fe^{+2} + H_2S \rightarrow FeS \text{ (Insoluble)}$$

The H_2S toxicity to rice may occur where Fe^{+2} is not found due to organo-iron complexation. The anaerobic bacteria (*Desulfovibrio* and *Desulfotomaculum*) can reduce sulphate during this respiration. Thus , under the submergence of soil , there is formation of insoluble metal sulphides, the Zn and Cu may be less available due to the formation of insoluble sulphides and H_2S toxicity may be enhanced due to low Fe^{+2} in soil.

Manganese

In submerged soils, there is reduction of manganese (Mn^{+4}) oxides to Mn^{+2} followed by a decrease due to precipitation of Mn^{+2} as $MnCO_3$ and reoxidation of Mn^{+2} to Mn^{+4} after diffusion to the oxidised soil-water – interface. The transformation of Mn^{+2} in submergence depends on pH, organic matter and active Mn content. The soil of low pH having high Mn and organic matter may build up Mn^{+2} (water soluble) nearing 90ppm with two weeks of submergence and then there is decline to a stable value(~10ppm). In alkaline soils, the concentration of Mn^{+2} hardly exceeds 10ppm under submergence. The concentration of water-soluble Mn after the peak may be explained by the equation (Eq. 9.22 and Eq.9.23) as:

$$pH + \frac{1}{2} \log Mn^{+2} + \frac{1}{2} \log pCO_2 = 4.4 \quad (9.22)$$

$$MnO_2 + 4\,H^+ + 2e^- \text{(reduction)} \rightarrow Mn^{2+} + 2H_2O \quad (9.23)$$

(Insoluble) (Water soluble, available to plant)

Iron

In waterlogged conditions, Fe^{3+} is reduced to Fe^{2+} with concomitant increase in Fe solubility by anaerobic bacteria, which uses iron oxides as electron acceptors in respiration. It produces toxic effect in rice plants called 'bronzing disease'.

Under anaerobic conditions, owing to the reducing process, hydrous iron oxides give rise to Fe^{2+} according to the following equation (Eq.9.24)

$$Fe(OH)_3 + e^- + 3H^+ \longrightarrow Fe^{2+} + 3H_2O \quad (9.24)$$

As a result of reduction of iron ,there is increase in the concentration of water-soluble iron, increase in solubility of P and Si. The rate of reduction of Fe^{+3} to Fe^{+2} depends on the length of flooding period, active manganese, nitrate, organic matter content etc. There is initial increase of Fe concentration in acid soil, rich in Fe on submergence to the tune of 600ppm(within 3 weeks) and thereafter , a sharp decline to a level of 50-100 ppm . The organic matter rich soil with low Fe content will have higher Fe concentration on submergence, while water soluble Fe rarely exceeds 20ppm in neutral and calcareous soils. The quantitative increase in water-soluble iron concentration can be explained by the following equation (Eq.9.25)

$$Eh = 1.058 - 0.059 \log Fe^{+2} - 0.177pH \quad (9.25)$$

The decrease in water soluble Fe concentration may be caused by the ppt of $FeCO_3$,$Fe_3(OH)_8$ or $Fe_3O_4.nH_2O$ after the decline of pCO_2 . However, the stability of Fe-content is regulated by the equation (Eq. 8.26):

$$Eh = 1.373 - 0.0885 \log Fe^{+2} - 0.236pH \quad (9.26)$$

After that, the water-soluble iron is deposited as iron mottles when diffused to the oxidised soil-water interface.

Zinc

Zn deficiency in flooded soils is thought to be related to high pH or the presence of $CaCO_3$. However, Zn deficiency also occurs in acid soils. The solubility of Zn is also affected by the reduction of hydrous oxides of Fe^{+3} and Mn^{+4}, changes in soil pH, increase in pCO_2 , insoluble sulphide and hydroxides and organo-metal complex. Following reactions may occur on soil submergence affecting the availability of Zn.

i) $Zn^{+2} + 2Fe^{+3} + 4H_2O$ (Reduction) $(Zn,Fe^{+2})Fe_2O_4$ [Franklinite]

ii) $Zn^{+2} + S^{-2}$ (Reduction) ZnS[Sphalerite]

iii) $Zn^{+2} + CO_2$ (High pCO_2) $ZnCO_3$[Smithsonite]

iv) $Zn^{+2} + 2OH^- \rightarrow Zn(OH)_2$

v) Adsorption of Zn^{+2} with inorganic and organic colloids.

vi) $Zn^{+2} + PO_4^{3-}$ (High P-fertilization) $Zn_3(PO_4)_2$

Copper: The solubility of Cu is regulated by the changes in soil pH, formation of insoluble sulphide and hydroxide, formation of organo-metal complexing agents and increase in pCO_2 on soil submergence. Due to the submergence, the availability of Cu in acid soil decreases for the formation of CuS,$CuCO_3$,$Cu(OH)_2$ as precipitation , whereas, it (Cu) increases markedly in alkali and calcareous soils.

$Cu_2(OH)_2CO_3$ (Malchite) + $4H^+ = 2Cu^{+2}$ (Soluble)+ CO_2 +$3H_2O$

Molybdenum and Boron: The availability of Mo is increased initially in acid soils on submergence and thereafter, remains almost static at the later stages. The adsorbed MoO_4^{2-} ions are released from the hydrous oxides of Fe and Mn with the increase in soil pH on submergence. The concentration of boron in soil solution on waterlogged soil remains almost unchanged.

Silicon: It has been observed that, the concentration of silica in the soil solution increases slightly on submergence and then there is gradual decrease . After the prolonged submergence in soil , the concentration of silica may be lower than the initial. The increase in concentration after flooding may be due to the mobilization of silica into the solution on reduction of hydrous oxides of Fe^{+3} sorbing silica or may be by the action of CO_2 on aluminosilicates . The decrease in the silica concentration at the later stages may be as a result of recommendation with aluminosilicates , following the decrease in pCO_2.

9.5 Management of Rice Soils

Rice is grown under well-puddled soil in a submerged condition. Transplanting or sowing of rice seedlings 2-3 weeks after submergence are practiced to provide a favourable environment to the roots. There should be facilities to drain out water once or twice in the early part of the growing season of rice for sufficient aeration. It requires to control the redox potential of the soil to avoid the toxic effects of Fe or H_2S on rice. A balance dose of fertilization is required for rice to maximise the yield. The addition of organic matter in soil may be done based on the physico-chemical condition of the soil and as per the requirement of the crop.

Questions

Objective

1. The 1st compound that undergoes reduction under submerged soil is
   a) Mn^{+4} b) SO_4^{2-}
   c) CO_2 d) NO_3
2. Soil submergence increases the specific conductance (True/False)
3. Eh of oxidised zone in submerged soil is
   a) > +400 mv b) > +350 mv
   c) >+300 mv d) > +250 mv
4. On account of waterlogged condition,
   a) The pH of acid soil increases
   b) The pH of alkali soil decreases
   c) The pH of alkali soil increases
   d) The pH of acid soil decreases
   e) Both a and b.
5. Due to submergence, H_2S toxicity occurs in
   a) Highly oxidised state b) Highly reduced state
   c) Highly reduced toxic state d) Both b and c
6. Soil submergence increases the availability of
   a) N,P and Mo b) N,P,Si and Fe
   c) N,P, Mn and Mo d) All
7. Due to submergence Fe^{+2} toxicity is found in
   a) Highly oxidised state b) Healthy reduced state
   c) Highly reduced toxic state d) Both b and c

8. Submergence of soil decreases the availability of
   a) Ca, Mg and S b) S, Cu, Zn
   c) Ca, Cu, Zn d) All
9. Critical limits of Eh for Fe^{+3} to Fe^{+2} is
   a) +280 to +220 mv b) +180 to +150 mv
   c) -120 to -180 mv d) -200 to -280 mv
10. The relation between pE and Eh are
   a) pE = Eh/0.0493 b) pE = Eh/0.0563
   c) pE = Eh/0.189 d) pE = Eh/0.0591
11. Submergence of soil causes
   a) NH_4 accumulation b) NH_3 volatilization
   c) Denitrification d) All
12. The concentration of B on soil submergence will
   a) Remain more or less constant b) Be increased
   c) Be decreased d) None
13. The availability of P under submergence is increased due to
   a) Increased solubility of strengite or variscite
   b) Mineralization of organic residues
   c) Increased solubility of Ca-PO_4
   d) Reduction of Fe-PO_4
   e) All
14. Due to submergence, the specific conductance of soil will be
   a) Decreased b) Increased
   c) Increased first and then decreased d) Zero
15. Critical limits of Eh for SO_4^{-2} to S^{-2} is
   a) +280 to +220 mv b) +180 to +150 mv
   c) -120 to -180 mv d) -200 to -280 mv

Subjective

1. What is submerged soil ?
2. Outline the sequential reduction of ions of a redox system in submerged soil.
3. What is Ferrolysis ?

4. What are the different forms of N, P and S under submerged soil ?
5. What do you mean by redox system in soil ?
6. Explain the equation : $MnO_2 + 4H^+ + 2e^- \rightarrow Mn^{+2} + 2H_2O$
7. What is the change of pH under water logged soil ?
8. What is the characteristics of a puddled soil ?
9. What is micro aerophilic zone under submerged soil ?
10. Explain the equation : $\Delta G = \Delta G^0 + RT \ln (Red)/(Ox)$

Descriptive

1. Outline in brief the types of submerged soil.
2. Describe the characteristics of submerged soil.
3. Outline the process of the Ferrolysis under waterlogged situation
4. Describe in brief the chemical transformation of N_2
5. Describe in brief the chemical transformation of P
6. Describe in brief the chemical transformation of micronutrients.
7. Describe the sequential reduction in waterlogged soil.
8. What is the fate of soil organic matter or added organic matter under water logged soil.
9. How does rice plant survive under waterlogged soil ?
10. What are the management options of submerged soil for yield maximization of crops ?

Suggested Readings

Ponnamperuma, F.N. (1972). The Chemistry of Submerged Soils. Advances in Agronomy. 24: 29-96.

Das, D.K. (1996). Introductory Soil Science . Kalyani Publishers.

10

Chemistry of Organo-Metal Complexation

The native and applied organic matter play an important role in terms of nutrient adsorption , nutrient mobility and microbial habitat on soil. The dominant ions present in soil under a particular agro-ecological situation may or may not be available to the growing plants due to the presence of organic matter. The binding energy and the stability of the organo-metal complexation would play an important role in availability of nutrients to the plants. The organo-metal complexation reaction depends on the nature of the ionic species (ionic/non-ionic) present in soil. The organic fractions (Humic acid/Fulvic acid) react with metal ions present in soil to form complexes which may be stable or unstable depending upon the molecular arrangement and degree of coiling of humic/fulvic substances. The surface charge density on clay and soil pH also play an important role to determine the order of stability of different nutrients (macro/micro) in soil.

10.1 Soil Organic Matter

Soil organic matter is one of the important constituents of soil which contains carbon, the most dominant element in the biological world. Organic matter consists mainly of the living organisms in the soil , the residue , and decomposed materials (Figure 10.1). Soil organic matter (SOM) is formed through the partial decomposition and transformation of plant inputs (above- and below-ground materials) by soil organisms. It includes living plant roots and animals, plant and animal remain at various stages of decomposition, and microorganisms and their excretions. The main sources of organic matter are plant litter (plant roots, stubble, leaves, mulch) and animal manures in the farm.

10.1.1 Functions of Organic Matter

Organic matter serves two main functions:

i) As soil organic matter is derived mainly from plant residues, it contains all of the essential plant nutrients. Hence, accumulated organic matter may be considered as a storehouse of plant nutrients.

ii) The stable organic fraction (humus) can adsorb nutrients in a plant-available form.

Organic matter releases nutrients in a plant-available form upon decomposition. In order to maintain the balance in nutrient cycling system, the rate of organic matter addition from crop residues, manure and any other sources must equal the rate of decomposition. The rate of uptake by plants and losses by leaching and erosion need to be addressed to ascertain the carbon stock in the environment.

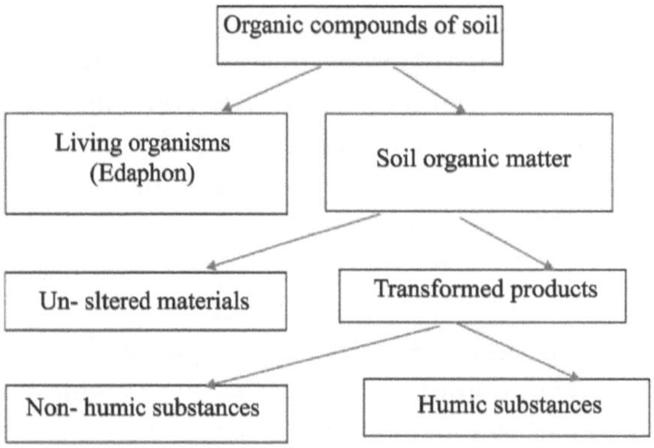

Figure 10.1. Composition of organic compounds in soil

10.1.2 Metal Ions

A metal ion is a type of atom in a compound possessing an electric charge. Such atoms when lose electrons to build positive ions called cations. Sometimes, ions are essentially enclosed by delocalized electrons and get involved in the processes like conductivity. Metal ions are formed by the removal of one or more outer electrons from the neutral atom. The energy required for ion formation is derived from the environment in which the metal exists. A considerable amount of energy is required for the formation of ions in gaseous phases. The electrolytes like sodium, chloride, potassium, calcium and magnesium, the nutritional elements when are dissolved in water get separated into positive and negative ions. As we know that, a metal is made up of atoms that primarily take part in the reaction processes to form different products still being attached in the structure, where the coordination number (number of anions surrounding the central cations) play an important role to form different crystals on earth.

10.2. Soil Organic Matter Fractions

Soil Organic Matter (SOM) on fractionation (Figure 10.2) results in humic and non-humic substances. The humic substances on sequential extraction produces humic acid,fulvic acid and humin. These fractions are the non-descriptive mixture of different chemical compounds, for which, humic acid, fulvic acid so extracted can not be considered as distinct chemical entities, as the isolates by precipitation (on adjusting the pH in the mixture) is based on the ability to form H-bonds. Each of the fractions of SOM has beneficial attributes, and these attributes are dependent on soil type.

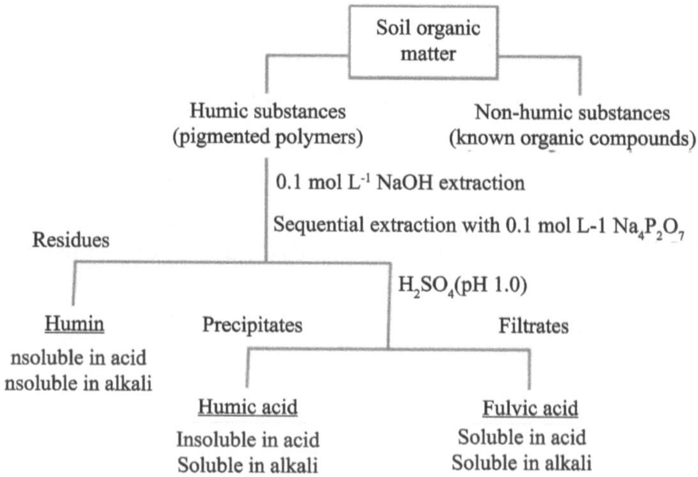

Figure 10.2. Fractionation scheme of soil organic matter

10.3. Complexation Reactions

A complexation reaction can be described as a reaction that forms a "complex". For example, when a cobalt salt, [$CoCl_{2\,(s)}$] is added to water, $[Co(H_2O)_6]^{2+}{}_{(aq)}$ is formed. The complexation reaction involves the combination of a metal ion with a compound containing chelate-forming, electron-donating functional groups to form metal chelates. Metal complexes can be formed with a central metal atom or ion surrounded by several atoms, ions or molecules, called ligands. Ligands are ions or molecules that can have an independent existence, and are attached to the central metal atom or ion. Examples of ligands are halide ions, carbon monoxide, ammonia, cyanide ion, etc. It has been noted that, a complex ion may be formed from a metal ion and a ligand because of a Lewis acid–base interaction. The positively charged metal ion acts as a Lewis acid, and the ligand, with one or more lone pairs of electrons, acts as a Lewis base.

10.4. Stability Constants

In coordination chemistry, a stability constant (also called formation constant or binding constant) is an equilibrium constant for the formation of a complex in solution. It is a measure of the strength of the interaction between the reagents that come together to form the complex. The stability constant is applicable to measure the strength of interactions between the ligands and metal ions that are involved in complex formation in the solution . The reaction can be expressed as the followings :

$$\text{Metal+Ligand} \leftrightarrows \text{Metal–Ligand}$$

$$K1=(ML)/[M][L]$$

Where, M = Metal;L= Ligand;ML = Metal-Ligand complex ;K1 = Constant

If the concentrations of the transition element complex and the reacting ligands are known, the expression for the stability constant (K_{stab}) can be used to determine which complex is more stable. The greater the value of K_{stab} the more stable the complex is.

The solubility of metal-humic acid complexes is expressed as stability constant.

If M = divalent metal ion that reacts with humic acid (HA) forming a metal -humic acid complex (MA_2)

Hence, $M^{2+} + 2HA \leftrightarrow MA_2 + 2H^+$

Then the equilibrium constant (K) will be as per the law of mass action:

$$K = [\ (MA^2)\ (H^+)^2]/[(M^{2+})\ (HA)^2]$$

Taking logarithm on both sides,

$$\log K = \log\ [(MA^2)(H^+)^2]/[(M^{2+})(HA)^2]$$

if the activity of HA and MA^2 are considered unity, then;

$$\log K = \log(H^+)^2\ /(M^{2+})$$

or,

$$\log K = 2\log(H^+)-\log\ (M^{2+})$$

where, logK is called the stability constant.

It was observed that, higher the value of the logK, higher is the stability of the complex compound . The logK also depends on the pH, ionic strength and concentration of reactants.

During the formation of complex compound, each radical is linked by only one coordination bond to the central metal ion.

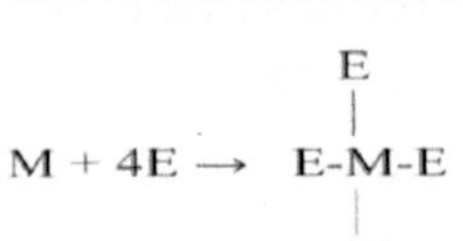

While, in chelates, a radical in the molecule is linked to the central metal ion by more than one coordination bonds.

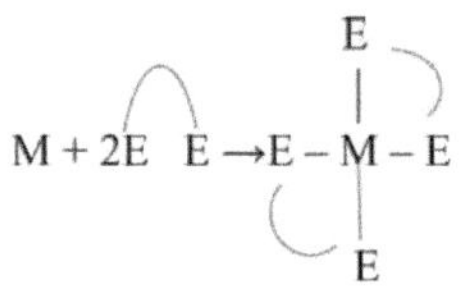

where, M = Metal, E = Radical

In a separate experiment it was observed (Mukhopadhyay and Sanyal 2004) that, the stability constant (log*K*) of the complexes formed by the natural humic/fulvic and synthetic humic acids with arsenic in aqueous phase evaluated by the ion-exchange method were quite stable, indicating their dependence on the nature and properties of the humic polymer, which, in turn, would affect the retention/release of such toxic metal in soil matrix.

10.5. Clay - Organic Interactions

Humus, the organic amorphous colloid supplies both basic and acidic ions which is transitory and disappear from soil later. On the other hand, the clay, the inorganic crystalline colloid supplying the basic nutrient ions is more or less stable. Both these colloids (Humus and clays) form the soil colloidal complex (Figure 10.3) and are active to facilitate plant nutrients.

It has been reported that, various clay minerals can bind humus substances through linkages known as clay-humus complex. Many investigators consider that the complexes between humus substances and clays (Figure 10.3) are probably formed by bridging through the exchangeable cations like Ca^{+2}, Mg^{+2} and Al^{+3}. Besides, humus substances can penetrate into the inter layer space of the crystalline clay mineral lattice.Another mechanism for the formation of stable clay-humus complexes are associated with the help of sesquioxides, which make specific bridges between humus substances and the crystalline lattice of clay minerals. Presumably these bridges are complex Al and Fe-humus compounds with non-silicate forms of sesquioxides. They are fixed to the surface of the clay minerals during dehydration by a process of adhesion at the expense of inter molecular types of linkages. There are two types of linkage taking place during the interaction of clay with humic acids:

i) Unstable-type of linkage: Humic acid is arranged on the external micelle surface of the clay and is easily extracted by alkali.

ii) Stable-type of linkage : This type of linkage is formed from the interaction of clay, humic acid and the exchangeable cations (Fe^{3+}, Al^{3+}, Ca^{2+} and Mg^{2+}) occurring in the lattice of the clay minerals. The phenolic -OH and COOH groups in the humic acid also take part in the linkage.

The formation of clay-humus complexes has some practical implications for the establishment of soil structure and for fixing organic substances in the soil.

Clay - Mineral Fractions + Organic Fractions = Clay-Humus complex

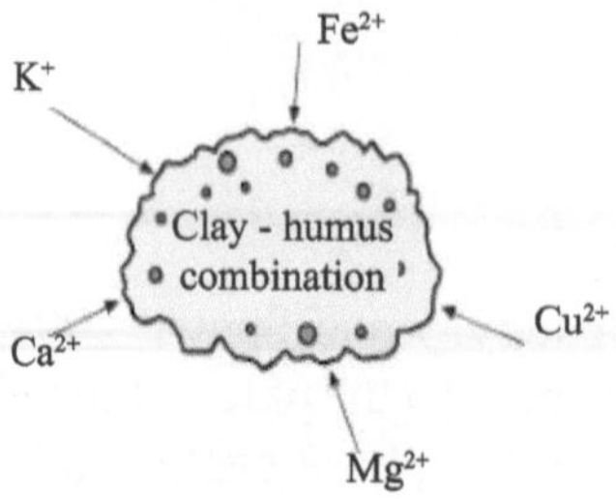

Figure 10.3. Formation of clay - humus complex

Besides, the clay minerals are known to have propensity to react with organic molecules and can catalyse various organic reactions due to its fine particle size, large surface area, layer structure, and charge characteristics. Adsorption on the different surfaces of a clay mineral lowers the dimensionality of the reaction space from 3 to 2. This restriction results in greatly enhanced reaction rates. Several Diels–Alder cycloadditions are catalyzed by Lewis acids and are sensitive to clay mineral catalysts (Figure 10.4).

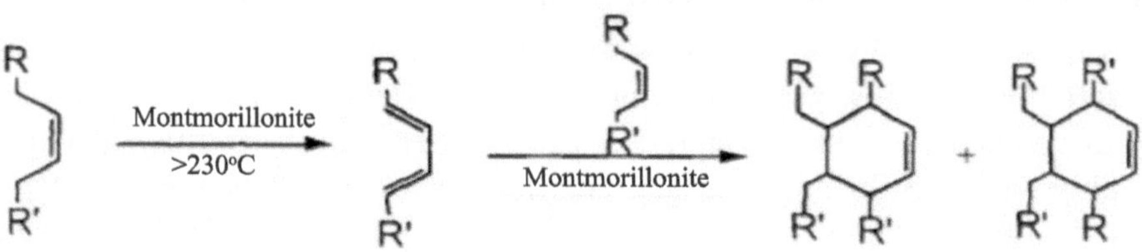

Figure 10.4. Clay - Organo interaction

An early Diels–Alder reaction is the dimerization of oleic acid on montmorillonite. (It is presumed that only the first step of dehydrogenation is catalyzed by montmorillonite, but not the Diels–Alder addition part itself). Another important organic reaction is formed by the Lewis acid–catalyzed Friedel–Crafts reaction (Figure 10.5).

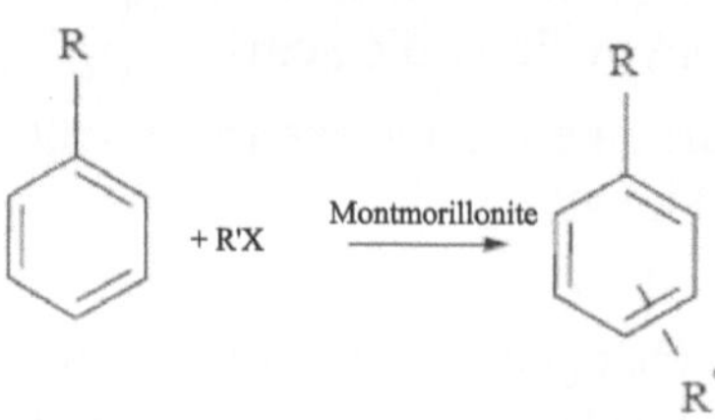

Figure 10.5. Montmorillonite-supported transition-metal salts (Zinc and Nickel chlorides)

Clay reacts with organic compounds to form complexes. The organic compounds may be anionic and polar non-ionic compounds.

10.5.1 Complex Formation with Organic Anions

Due to the negative charge of clay , there is possibility of repulsion with organic anions, although, a positive charge may develop depending upon the pH or zero point charge (ZPC) in the complex. There are two types of mechanisms of clay-organic anion interactions, namely, positive and negative adsorption. The other special conditions are metal and water – bridging, Vander Waals force and ligand exchange mechanisms.

a) **Positive and negative adsorption:** Positive adsorption is the attraction of organic anions by positively charged surface, developed under acidic condition at the surfaces. The acidic species in organic compounds persist at soil pH below its pKa value and behaves as cations and thus will be adsorbed by the negatively charged clay surfaces.

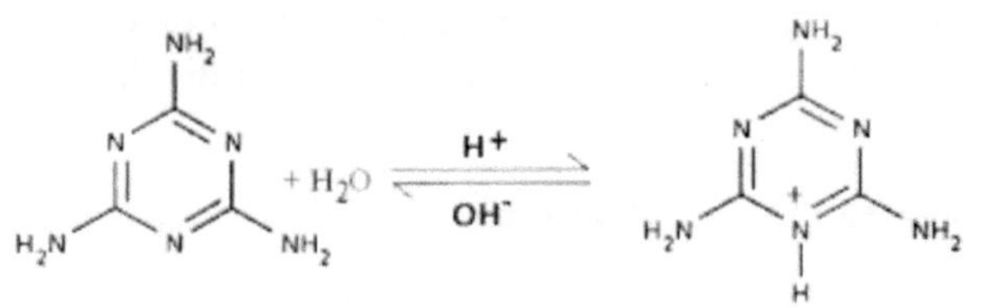

Melamine (deprotonated form) Melamine (protonated form)

Figure 10.6. Conversion of melamine from deprotonated to protonated form

However, anionic forms of the organic compound will exist at soil pH, above its pKa value. The transformation into a cationic forms from the basic / neutral species may occur due to protonation reaction (Figure 10.6).

The negative adsorption has been reported for the Herbicides like 2,4-D and 2,4,5-T by Na-Smectite in a medium with a pH above the pKa value of the organic compound and it continues until the pH of the medium equals the pKa, or the dissociation constant of the respective compound. When the pH decreases below the pKa, then positive adsorption will start. The organic substances are found in basic forms at pH values greater than pKa to behave as anions and thus become accessible for attraction by the positively charged ions or be rejected for repulsion by negatively charged clay.

b) **Metal and water bridging:** The phenomenon of metal bridging between organic anions and clay minerals exhibits the permanent negative charge (e.g. 2:1 type clays). The presence of polyvalent cations which can maintain neutrality at the interface for neutralising negative charges of clays or organic complexes would make such type of interaction. As per the proposition (Stevenson,1994),

i) The water molecules present in the hydration shell of the polyvalent cations are not displaced or react with organic anions by hydrogen bonding.

ii) The water molecules displaced from the hydration shell of the polyvalent cations and anions are coordinated directly to the cations.

iii) Van der Waals force : This force can provide weak force of attraction by close contact and is operational with all molecules. The Van der Waals force is followed by coordination bonding for the formation of complex or for chelation reaction. The Vander Waals force is very much important in adsorption of neutral polar and non-polar organic compound (Stevenson,1994).

iv) The interaction of fulvic acid with Humus- rich allophanic clays in Andosols is mediated by Ligand exchange mechanisms (Tan,1982). During the interaction, carboxylic group of fulvic acid can displace the OH group present on the allophane (Parfitt *et al.*,1977).

10.5.2 Complex Formation with Organic Cations

Presence of positively charged organic cations will attract the negatively charged clay minerals due to the intimate contact under normal condition.

On acceptance of protons, the organic compound becomes positively charged as followings:

a) Donation of proton by the H^+- saturated clay.

$R\text{-}NH_2 + H\text{-clay} \leftrightarrow R\text{-}NH_3{}^{(+)} - \text{clay}$

b) Donation of proton by the polarised water

$(M.mH_2O)^{m+} + NH_2\text{-}R \leftrightarrow RNH_3{}^+ + [\,M.(m\text{-}1)H_2O\,]^{(m-1)+}$

Where, M= cation with m+ charges

10.5.3 Complex Formation with Amphoteric Organo-Compounds

Sometimes the presence of carboxylic and amino groups in humic compounds provide the ability to exist the amino acids either as cations, anions or as Zwitterions (Figure 10.7). The mechanism of proton transfer are the main processes of clay -solution interface interaction.

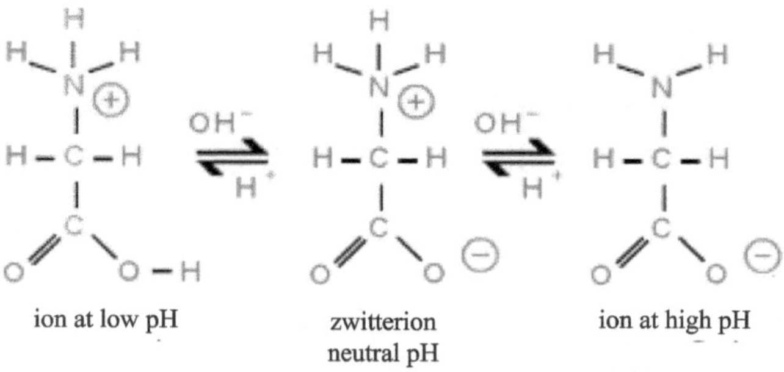

Figure 10.7. Conversion of amino acids into ionic species

The Zwitterion will act through ion dipole interaction, while the positive pole with negatively charged clay surfaces and the negative pole can react with positively charged cations during the metal ion interactions. Sometimes the anionic forms are attracted on the negatively charged clay surface by cationic or through water- bridging action (Figure 10.8).

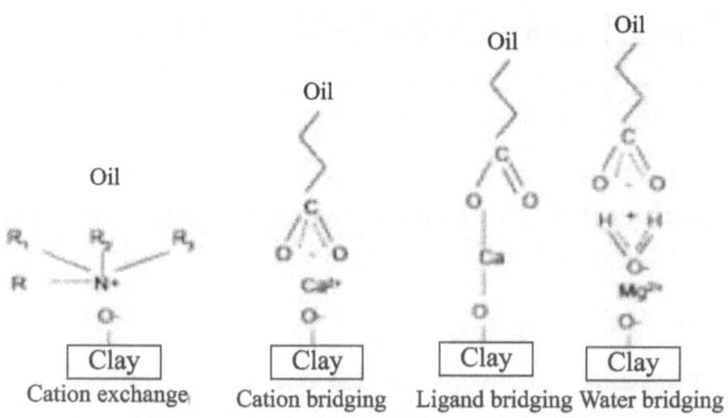

Figure 10.8. Water bridging by water molecule in the hydration shell

10.5.4 Complex formation with Non-ionic Organic Compounds

The non-ionic organic compounds are adsorbed at the surface of clay minerals or in the intermicellar space of expanding type of clays. The uncharged organic compound are adsorbed by van der Waals attraction and hydrogen bonding. The uncharged organic molecules (alcohol and ethylene glycol) could have interaction with clay in presence of exchangeable cations, which are surrounded by water molecules. The intercalation (adsorption of organic molecules in the intermicellar spaces) may occur during the adsorption of organic molecules which results increasing the basal spacing of clays.

10.5.5 Colour Change on Clay-organic Reactions

There may be changes of colour due to the reactions between organic compounds and clays. The clay-organic reactions was extensively reviewed

(Theng,1971) and clay- organic reactions involving colour changes was reported (Mielenz, King and Schlieltz 1950). It was found that clays could adsorb organic substances, either in a natural condition or pre- treated (usually with an acid, like HCl) , with an accompanying colour change which may be specific for a given species of clay mineral. Clays with a low cation exchange capacity (CEC), such as kaolinite, react slightly and cause a lesser colour change, whereas, clays with a high CEC, like the smectites, may cause a pronounced colour change. Some organic compounds are adsorbed by the clay without any change in colour. The colour change caused by clays treated with organic staining agents is believed to be due to either a redox reaction, generally because of iron in the clay, or because of an acid- base type of reaction (Hendricks and Alexander 1940; Hauser and Leggett 1940; Furukawa and Brindley 1973) .For example, yellow- brown, orange-brown , red- brown, yellowish to brownish-red, would be matched by the colours of goethite, lepidocrocite, and hematite. In other cases, yellow-green , greenish-yellow, greenish- brown or green colours would be very similar to those of the common ferromagnesian primary minerals. For this reason very few of the reagents can be used effectively for quantitative determination of clays in thin sections of rock .Methylene blue and benzidine are useful for quantitative determination of clay minerals in basaltic rocks because they produce a blue-green or blue colour in clays which is not matched by any blue- green primary or secondary minerals if found in basaltic rocks.

10.6 Soil Fertility and Complex Formation

We know that, presence of adequate organic matter in soil is good for maintaining soil structure, bulk density and water holding capacity of soil. Due to the complex formation between the clay and organic matter, a stable structure in soil is formed which governs the aggregation of soil particles and electrochemical properties of clay surfaces affecting water absorption. The polysaccharides becomes stable against microbial decomposition due to the complexation reactions in soils. The clay -mucigel interaction also help initiating the aggregation. Besides, the flocculated clay-humus complex is also required for the aggregation processes. Complexation reaction sometimes enhances the bio- availability of some of the essential metallic ions (e.g., micronutrient cations) which would have been unavailable from the soil if precipitated earlier. Hence, formation of clay-humus or organo-metal complexes could play a significant role in sustaining the soil fertility for a specific agro-ecological region.

Questions

Objective

1. Humic acid is

   a) Soluble in alkali
   b) Soluble in acid
   c) Soluble in both acid and alkali
   d) None of the above

2. Fulvic acid is

   a) Soluble in acid and alkali
   b) Insoluble in acid and alkali
   c) Soluble in acid
   d) Soluble in alkali

3. Humin is

   a) Insoluble in acid and alkali
   b) Soluble in acid and alkali
   c) Insoluble in acid
   d) Insoluble in alkali

4. The interaction of Fulvic acid with humus rich allophanic clay is mediated by

   a) Cation exchange reaction
   b) Anion exchange reaction
   c) Ligand exchange reaction
   d) None of the above

5. At low pH, there is development of

   a) +ve charge
   b) -ve charge
   c) 0 charge
   d) Both +ve and -ve charge from the Zwitterion

6. Uncharged organic compounds are attracted with clay by

   a) Cation bridge
   b) Anion bridge
   c) Vander Waals force
   d) None of the above

7. If the value of the stability constant (logK) between the metal-chelating compound is more

   a) More is the binding force
   b) Less is the binding force
   c) No binding force
   d) None of the above

8. The order of participation of groups in metal – humic complex are :

   a) -OH, -COOH, >C=O, >NH
   b) -COOH, -OH, >C=O, >NH
   c) >C=O, >NH, >OH-, -COOH
   d) > NH, - COOH, -OH, > C=O

9. The E_4/E_6 provides
   a) The relative preponderance of aliphatic and aromatic group
   b) Only the presence of aliphatic group
   c) Only the presence of aromatic group
   d) None of the above
10. The number of free radical contents of humic acids obtained from electron spin resonance (ESR) data decreases in the order:
    a) Humin > humic acid > fulvic acid
    b) Humic acid > fulvic acid > humin
    c) Fulvic acid > humin > humic acid
    d) None of the above
11. The beneficial effect of organic matter is very high in
    a) Fine textured soil b) Medium textured soil
    c) Coarse textured soil d) All
12. Precipitation of humic acid takes place in
    a) Dilute acid b) Dilute alkali
    c) Concentrated acid d) Concentrated alkali
13. Humin can be extracted by
    a) 2N HCl b) 2N NaOH
    c) 2N Na_2CO_3 d) 4% H_2O_2
14. Of the followings, which is/are the more stable metal-organic interactions found in soil ?
    a) Metal-humic acid b) Metal-fulvic acid
    c) Metal-humin d) Both a and b
    e) All the above
15. Polymerization of phenols in presence of formaldehyde will yield which of the followings ?
    a) Anion exchange resin b) Cation exchange resin
    c) Both a and b d) None

Subjective

1. What is organic matter -metal association in soil ?
2. Outline two major functions of soil organic matter.
3. What are the composition of soil organic matter ?

4. What are the fractions of soil organic matter ?
5. What is stability constant (logK)?
6. What is the effect of pH on Zwitterion ?
7. What happens when there is donation of proton by the polarised water ?
8. What are the extractants for humic acid fractions of soil ?
9. What is chelation reaction ?
10. What do you mean by water-bridging in clay organic complexation ?

Descriptive

1. Outline in brief the concept of organo-metal complexation reaction.
2. Outline the fractionation scheme of Soil organic matter.
3. Briefly describe the nature of complexation reaction in soil.
4. How stability constant (logK) of a metal -ligand reaction is derived ?
5. How does complex formation with organic anions held ?
6. How does complex formation with organic cations are obtained ?
7. Outline the complex formation with amphoteric organic compounds.
8. Outline complex formation with non-ionic organic compounds.
9. How does molecular weight of humic substances is determined by viscosity measurement.
10. What are the implications of complex formation in regard to soil fertility ?

Suggested Readings

Furukawa, T. and Brindley, G.W. (1973). Adsorption and oxidation of benzidine and aniline by montmorillonite and hectorite. Clays and Clay Minerals 21: 279-288.

Hauser, E.A. and Leggett, M.B. (1940). Colour reactions between clays and amines. J. Am. Chem. Soc. 62: 1811-1814.

Hendricks, S.B. and Alexander, L.T. (1940). A qualitative colour test for the montmorillonite type of clay minerals. J. Am. Soc. Agron. 32: 455-458.

Kim H.Tan (1982). Principles of Soil Chemistry. CRC Press. Taylor and Francis Group.

Mielenz, R.C., King, M.E. and Schlieltz, N.D. (1950). Staining tests. In Section 6, Analytical data on reference clay minerals. Am. Petrol. Inst. Proj. 49, Clay Mineral Standards, Prelim. Report No. 7: 135-160.

Mukhopadhyay, D. and Sanyal S.K. (2004). Complexation and release isotherm of arsenic in arsenic-humic/fulvic equilibrium study. *Australian Journal of Soil Research*, 42: 815-824.

Stevenson, F.J. (1994). Humas chemistry, genesis, composition, reactions, 2nd ed. Wiley and Sons, New York.

Theng, B.K.G. (1971). Mechanisms of formation of colored clay-organic complexes. A review. Clays and Clay Minerals 19: 383-390.

Appendix

Appendix A
The Greek Alphabets

Α Alpha al-Eab	Β Beta bay-tah	Γ Gamma gam-mah	Δ Delta del-tah	Ε Epsilon ep-si-lon	Ζ Zeta zay-tah
Η Eta ay-tah	Θ Theta thay-tah	Ι Iota eye-o-tah	Κ Kappa cap-ah	Λ Lamb da lam-dah	Μ Mu mew
Ν Nu new	Ξ Xi rr-eye	Ο Omicron om-e-cron	Π Pi pie	Ρ Rho roe	Σ Sigma sig-mah
Τ Tau taw	Υ Upsilon oop-si-lon	Φ Phi fie	Χ Chi k-eye	Ψ Psi sigh	Ω Omega o-may-gah

Appendix B
Prefixes for SI Units

SI Prefixes					
Number	**Prefix**	**Symbol**	**Number**	**Prefix**	**Symbol**
10^{1}	deka	da	10^{-1}	deci	d
10^{2}	hecto	h	10^{-2}	centi	c
10^{3}	kilo	k	10^{-3}	milli	m
10^{6}	mega	M	10^{-6}	micro	u (greek mu)
10^{9}	giga	G	10^{-9}	nano	n
10^{12}	tera	T	10^{-12}	pico	p
10^{15}	peta	P	10^{-15}	femto	f
10^{18}	exa	E	10^{-18}	atto	a
10^{21}	zeta	Z	10^{-21}	zepto	z
10^{24}	Yotta	Y	10^{-24}	yocto	y

Appendix C
Conversion factor

Mass & Weight
1 kg = 1000g = 2.205 pounds
1 gram = 1000 mg
1 pound = 453.59 grams
1 amu = 1.6606×10^{-24} grams
1 gram = 6.022×10^{23} amu
1 ton = 2000 pounds
16 ounces = 1 pound

Length
1 inch = 2.54 cm
1 meter = 100 cm = 39.7 inches
1 yard = 0.9144 meters
1 mile = 1.6093 kilometrs
1 kilometer = 1000 meters = 0.621371 miles
1 Angstrom = 1.0 x 10-10 meters = 1.0×10^{-8} cm
1 troy ounce = 31.3g

Volume
1 liter = 0.001 cubic meters (M^3)
1 liter = 1000 cm^3 = 1000 mL
1 liter = 1.056 quarts
1 quart = .94631 liter
1 mL = 0.001 liters = 1 cm^3
1 ft^3 = 7.475 gallons = 28.316 liters
1 cup = 8 ounces
1 quarts = 4 cups 2 pints
4 quarts = 1 gallon
1 ounce = 29.57 cm^3 or mL

Metric Conversions		
Unit	**Abbr.**	**Meaning (Example: 1 gram = 10^3 kg)**
Mega-	M	10^6
kilo-	k	10^3
hecto-	h	10^2
deca-	dk	10^1
deci-	d	10^{-1}
centi-	c	10^{-2}

mili-	m	10^{-3}
micro-	u	10^{-6}
nano-	n	10^{-9}

Conversion of plant nutrient

Column l	Multiply by **	Column 2
P2O5	0.437	P
K2O	0.830	K
Cao	0.715	ca

Appendix D

Periodic table of the elements

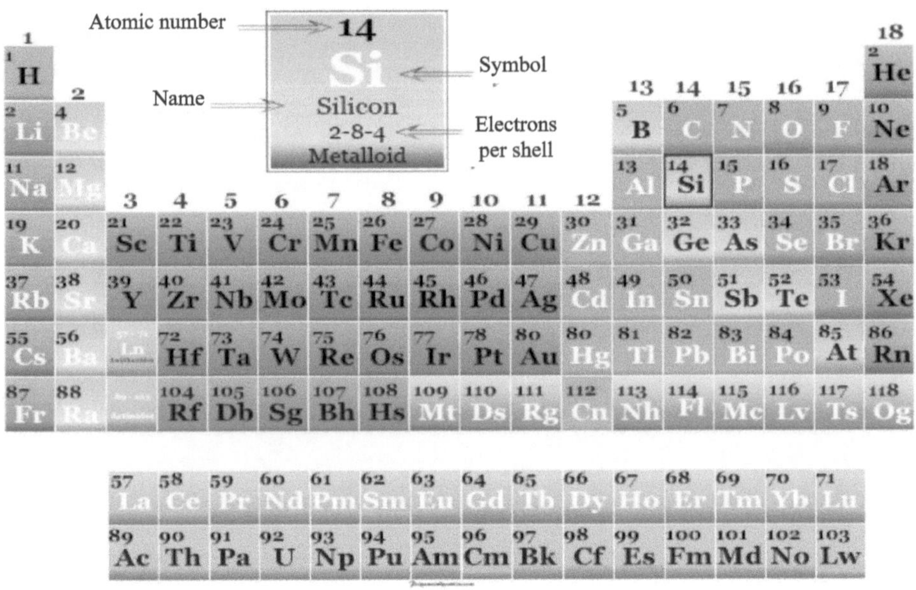

References

Adams, J.M, Dyer, S., Martin, K., Matear, W.A., McCabe, R.W. (1994). Diels–Alder reactions catalysed by cation-exchanged clay minerals. J. Chem. Soc. Perkin Trans. 1: 761-765.

Bear, F.E. (1964). Chemistry of the Soils. Oxford and IBH.

Beckett, P.H.T. (1964). Studies on soil potassium: 1. Confirmation of the ratio law: measurement of potassium potential. Journal of Soil Science, 15: 1-8.

Biswas, T.D. and Mukherjee, S.K. (1994). Textbook of Soil Science. Tata McGraw-Hill Publishing Company Limited, New Delhi.

Brady, N.C. and Weil, R.R. (2007). The Nature and Properties of Soils. 14th Edition, Prentice Hall, USA.

Das, D.K. (1996). Introductory Soil Science. Kalyani Publishers.

Datta, A., Sanyal, S.K. and Saha, S. (2001). A study on natural and synthetic humic acids and their complexing ability towards cadmium. Plant and Soil , 235: 115-125.

Davis, L.E. (1945). Simple kinetic theory of ionic exchange for ions of unequal charge. J. Phys. Chem. 49: 473-479.

Frederick K. L. and Edward J. T. (2000). Essentials of Geology. 7th Ed., Prentice Hall.

Furukawa, T. and Brindley, G.W. (1973). Adsorption and oxidation of benzidine and aniline by montmorillonite and hectorite. Clays and Clay Minerals 21: 279-288.

Gapon, E.N. (1933). Theory of exchange adsorption in soil. J. Gen. Chemistry (USSR). 3: 144-152.

Glasstone, S. (1946). Textbook of Physical Chemistry; Macmillan student editions.

Greenland, D.J. and Hayes, M.H.B. (1981). Chemistry of Soil Processes. John Wiley and Sons.

Grim, R.E. (1968). Clay Mineralogy. Mc Graq Hill.

Hauser, E.A. and Leggett, M.B. (1940). Colour reactions between clays and amines. J. Am. Chem. Soc. 62: 1811-1814.

Havlin, J. L., Tisdale, S.L., Beaton, J.D. and Nelson, W.L. (2007). Soil fertility and fertilizers. (Seventh Edition). First impression. Prentice Hall of India Ltd. New Delhi.

Haylock, O.F. (1956). A method for estimating the availability of non-exchangeable potassium. Proceedings of 6th International Congress of Soil Science. Part B: 403-408.

Hendricks, S.B. and Alexander, L.T. (1940). A qualitative colour test for the montmorillonite type of clay minerals. J. Am. Soc. Agron. 32: 455-458.

Jenny, H. (1936). Simple kinetic theory of ionic exchange. I. Ions of equal valency. J. Phys. Chemistry. 40: 501-517.

Jenny, H. (1941). Factors of Soil Formation, McGraw Hill Publ, New York.

Joffe, J.S. (1949). Pedology. New Brunswick, New Jersey, Pedology Publication.

Kaur, N. and Kishore, D. (1012). Montmorillonite: An efficient, heterogeneous and green catalyst for organic synthesis. J. Chem. Pharm. Res. 4: 991-1015.

Kim H.Tan (1982). Principles of Soil Chemistry. CRC press. Taylor and Francis Group.

Kleijn,W.B. and Oster, J.D. (1983). Effects of permanent charge on the electric double layer properties of clays and oxides. Soil Sci. Soc. Am. 47: 821-827.

Lindsay, W.L. (1979). Chemical equilibria in soils. New York; John Willey and Sons.

Losfeld, G., Escande, V., Vidal de la Blache, P., l'Huillier, L., Grison, C. (2012). Design and performance of supported Lewis acid catalysts derived from metal contaminated biomass for Friedel-Crafts alkylation and acylation. Catal. Today. 189: 111-116.

Manoj Kumar (1995). Ph.D. thesis IARI, New Delhi, cited in Ghosh and Ghosh (1998).

Marschner, H. (1995). Mineral nutrition of higher plants, 2nd Ed. (New York; Academic Press).

Marshall, C.E. (1949). The colloid chemistry of silicate minerals. Academic Press, New York.

Marshall, C.E. (1964). Physical chemistry and minerology of soils, Vol.1. John Wiley and Sons, New York.

Martin, D., Srivastava, P.C., Ghosh, D. and Zech, W. (1998). Characteristics of humic substances in cultivated and natural forest soils of Sikkim. Geoderma. 84: 345-362.

McLean, A.J. (1961). Potassium supplying power of some Canadian soils. Can. of Soil Sci. 59(30): 295-299.

Mielenz, R.C., King, M.E. and Schlieltz, N.D. (1950). Staining tests. In Section 6, Analytical data on reference clay minerals. Am. Petrol. Inst. Proj. 49, Clay Mineral Standards, Prelim. Report No. 7: 135-160.

Mohsin, M.A., Sarkar, A.K. and Mathur, B.S. (1995). Acid soil management. Kalyani Publishers.

Mortvedt, J.J., Cox, F.R., Shuman, L.M. and Welch, R.M. (1991). Micronutrients in Agriculture. 2nd Ed. SSSA Book Sewries No.4. (Madison, Wisconsin: Soil Sci. Soc. Amer.)

Mukhopadhyay, D. and Sanyal S.K. (2004). Complexation and release isotherm of arsenic in arsenic-humic/fulvic equilibrium study. Aus. of Soil Res., 42: 815-824.

N.K. Dey and Sarkar, H.K. (1993). Soil Geography. Shribhumi Publishing Company, Kolkata

Nagendrappa, G. (2002). Organic Synthesis using Clay Catalysts. Clays for 'Green Chemistry'. Resonance, 64-77.

Page,J.B and Baver, L.D. (1940). Soil Science Society of America Proceedings, 4,150-155.

Pal, S.K. (2016). Text book of soil science. Oxford and IBH Publishing Company Pvt. Ltd.

Palit, S.R. (1938). Elementary Physical Chemistry, Book Syndicate Pvt. Ltd., Calcutta.

Parfitt, R.L., Fraser, A.R. and Farmer, V.C. (1977). Adsorption of hydrous oxides. III. Fulvic acid and humic acid on goethite, gibbsite and imogolite. J. Soil Sci. 28: 289-296.

Ponnamperuma, F.N. (1972). The Chemistry of Submerged Soils. Adv. in Agron. 24: 29-96.

Rakshit, P.C. (2001). Physical Chemistry. Publisher, Sarat Book Distributors, ISBN

Raman, K.V. and Ghosh, S.K. (1974). Identification and quantification of minerals in clays. In: Minerology of Soil Clays and Clay Minerals, Bulletin No. 9 (S.K. Mukherjee and T.D. Biswas, Eds). Indian Soc. of Soil Sci., New Delhi, pp. 117-142.

Russel, E.J. and Appleyard, A. (1915). The atmosphere of the soil, its composition and causes of variation . J. Agr. Sci. 7: 1-48.

Sanyal, S.K. (1984). Structure of water solutions of organics- Hydrophobic hydration. Chem. Edn. (UGC), 1(2): 14-18.

Sanyal, S.K. (2002). Fundamentals of soil science. Published on behalf of the Indian Society of Science, IARI, New Delhi.

Sanyal, S.K. (2002). Soil colloids. In: 'Fundamentals of Soil Science' (G.S. Sekhon, P.K. Chhonkar, D.K. Das, N.N. Goswami, G. Narayanasamy, S.R. Poonia, R.K. Rattan and J.L. Sehgal, Eds.). Indian Society of Soil Science, New Delhi, pp. 229-259.

Sanyal, S.K. (2018). A text book of Soil Chemistry. Daya Publishing House, Division of Astral International Pvt. Ltd. New Delhi 110002.

Sanyal, S.K., Poonia, S.R. and Barua, T.C. (2009). Soil colloids and ion exchange in soil. In: Fundamentals of Soil Science. (N. N. Goswami, R. K. Rattan, G. Dev, G. Narayanaswamy, D. K. Das, S.K.Sanyal, D. K. Pal and D. L. N. Rao eds.). Second edition. Indian Society of Soil Science, New Delhi.

Schnitzer, M. and Khan, S.U. (1972). Humic substance in the environment, Marcel Dekker, New York.

Schofield, R.K. and Taylor, A.W. (1955). The measurement of soil pH. Soil Sci. Soc. Am. Proc. 19: 164-167.

Schwertmann, U. and Taylor, R.M. (1977). Iron oxides. In: Minerals in soil environments. J.B. Dixon, S.B. Weed, J.A. Kittrick, M.H. Milford and J.L. White (eds.). Soil Science Society America, Madison, WI, pp. 145-180.

Simonson, R.W. (1968). Concept of Soil. Adv. Agron. 20: 1-47.

Solomon, O.H., Loft, G.C. and Swift, A. (1968). Reactions catalysed by minerals-IV: The mechanism of the benzidine blue reaction. Clay Minerals. 7: 389-397.

Sposito, G. (1989). The Chemistry of Soils. Oxford University Press. NewYork, NY.

Sposito, G. (2008). The Chemistry of Soils. 2nd ed. Oxford University Press. NewYork, NY.

Stevenson, F.J. (1994). Humas chemistry, genesis, composition, reactions, 2nd ed. Wiley & Sons, New York.

Tan, K.H. (2005). Principles of Soil Chemistry. Marcel Dekker Inc., New York.

Theng, B.K.G. (1971). Mechanisms of formation of colored clay-organic complexes. A review. Clays and Clay Minerals. 19: 383-390.

Tombecz, E. (1999). Colloidal properties of humic acids and spontaneous changes of their colloidal state under variable solution conditions. Soil Sci. 164: 814-824.

Varadachari, C., Mondal, A.H., Nayak, D.C. and Ghosh, K. (1994). Clay humas complextaion. Effect of pH and the nature of bonding. Soil Biol. Bioche., 26: 1145-1149.

Visser, S.A. (1985). Viscometric studies on molecular weight fractions of fulvic acids and humic acids of aquatic terrestrial and microbial origin. Plant Soil. 87: 209-221.

Yetes, D.E., Levine, S. and Healy, T.W. (1974). Site binding model of the electric double layer at the oxide /water interface. J. Chem. Soc. Faraday Trans. I. 70: 1807-1819.

Yuan, G., Theng, B.K.G., Parfitt, R.L. and Percival, H.J. (2000). Interactions of allophane with humic acid and cations. European J. of Soil Sci., 51: 35-41.

Zaccheria, F., Santoro, F., Iftitah, E.D., Ravasio, N. (2018). Brønsted and Lewis solid acid catalysts in the valorization of citronellal. Catalysts. 8: 410.

Index